New Techniques in Digital Holography

Series Editor
Dominique Placko

New Techniques in Digital Holography

Edited by

Pascal Picart

iSTE

WILEY

First published 2015 in Great Britain and the United States by ISTE Ltd and John Wiley & Sons, Inc.

ISTE Ltd
27-37 St George's Road
London SW19 4EU
UK

www.iste.co.uk

John Wiley & Sons, Inc.
111 River Street
Hoboken, NJ 07030
USA

www.wiley.com

Library of Congress Control Number: 2014958258

British Library Cataloguing-in-Publication Data
A CIP record for this book is available from the British Library
ISBN 978-1-84821-773-7

Contents

Introduction

Holography, the brilliant idea from Dennis Gabor [GAB 48], became "digital" in the early 1970s with the pioneering works of Goodman, Hua and Kronrad [GOO 67, HUA 71, KRO 72]. It took until 1994, for "digital holography" based on array detectors to come about [SCH 94], as a consequence of the important developments in two sectors of technology: microtechnological procedures have made the creation of image sensors with numerous miniaturized pixels possible, and the rapid computational treatment of images has become accessible with the appearance of powerful processors and an increase in storage capacities. From 1994, holography found a new life in the considerable stimulation of research efforts. About 20 years later, digital holography appears to be a mature topic, covering a wide range of areas such as three-dimensional (3D) imaging and display systems, computer-generated holograms, integral imaging, compressive holography, digital phase microscopy, quantitative phase imaging, holographic lithography, metrology and profilometry, holographic remote sensing techniques or full-field tomography. In addition, besides the visible light classically used, light sources including coherent to incoherent and X-ray to terahertz waves can be considered. Thus, digital holography is a highly interdisciplinary subject with a wide domain of applications: biomedicine, biophotonics, nanomaterials, nanophotonics, and scientific and industrial metrologies.

Introduction written by Pascal PICART.

Thus, as actors of this boom, it seemed convenient that we propose a book devoted to special techniques in digital holography. The coauthors aim to establish a synthetic stat of the art of important advances in the field of digital holography. We are interested in detailing advances related to fundamentals of digital holography, in-line holography applied to particle tracking and sizing, digital color holography applied to fluid mechanics, digital holographic microscopy as new modality for live cell imaging and life science applications, long-wave infrared holography, and special techniques in full-field vibrometry with detection at the ultimate limits.

The book is organized into seven chapters. Chapter 1 introduces the basic fundamentals of digital holography, the recording of digital holograms, demodulation techniques to separate the diffraction orders, algorithms to reconstruct the complex object wave, and basic principles of holographic interferometry and phase tomography. Chapter 2 discusses the use of in-line holography for the study of seeded flows; the recent developments permit us to apply this technique in many industrial or laboratory situations for velocimetry, particle size measurement or trajectography. In Chapter 3, the coauthors present new approaches in three-color holography for analyzing unsteady flows. Special techniques to visualize and quantitatively analyze flows up to Mach 10 are presented. The in-line approach based on Wollaston prisms will be discussed and compared to the holographic Michelson arrangement. Chapter 4 is devoted to automation of digital holographic detection procedures for life sciences applications. With the use of partial spatially coherent light sources, the use of a reduced coherence source is of interest for reducing the measurement noise; typical applications are detailed. The coauthors describe specific tools linked to the numerical propagation that are indispensable to process the information correctly, avoid numerical effects and make easier the further processing. The automated 3D detection methods based on propagation matrices with both a local and a global approach are discussed and illustrated on concrete applications. Chapter 5 is devoted to applications of quantitative phase digital holographic microscopy in cell imaging. The most relevant applications in the field of cell biology are summarized. Recent promising applications obtained in the field of high content screening are presented. In addition, the important issue concerning the development of multimodal microscopy is addressed and illustrated trough concrete examples, including combination with fluorescence microscopy, Raman spectroscopy and electrophysiology. Chapter 6 presents digital holography in the long-wave infrared domain. Technology related to sensors and light sources are presented and digital holographic infrared interferometry is detailed and applied to high-amplitude displacements of industrial aeronautic structures. Examples of non-destructive testing (NDT) are also provided. Chapter 7 presents new techniques in the field of vibration measurement; combined off-axis and heterodyne digital holography experiments are presented. In particular, techniques based on high speed and ultimate sensitivity are described. Examples related to life sciences are presented and detailed.

This book is intended for engineers, researchers and science students at PhD and Master's degree level, and will supply them with the required basics for entering the fascinating domain of digital holography.

I.1. Bibliography

[GAB 48] GABOR D., "A new microscopic principle" *Nature*, vol. 161, pp. 777–778, 1948.

[GOO 67] GOODMAN J.W., LAWRENCE R.W., *Applied Physics Letters*, vol. 11, pp. 77–79, 1967.

[HUA 71] HUANG T.S., *Proceedings of the IEEE*, vol. 159, pp. 1335–1346, 1971.

[KRO 72] KRONROD M.A., MERZLYAKOV N.S., YAROSLAVSLII L.P., *Soviet Physics Technical Physics*, vol. 17, pp. 333–334, 1972.

[SCH 94] SCHNARS U., JUPTNER W., *Applied Optics*, vol. 33, pp. 179–181, 1994.

1

Basic Fundamentals of Digital Holography

The idea of digitally reconstructing the optical wavefront first appeared in the 1960s. The oldest study on the subject dates back to 1967 with the article published by Goodman in *Applied Physics Letters* [GOO 67]. The aim was to replace the "analog" recording/decoding of the object by a "digital" recording/decoding simulating diffraction from a digital grating consisting of the recorded image. Thus, holography became "digital", replacing the silvered support with a matrix of the discrete values of the hologram. Then, in 1971, Huang discussed the computer analysis of optical wavefronts and introduced for the first time the concept of "digital holography" [HUA 71]. The works presented in 1972 by Kronrod [KRO 72] historically constituted the first attempts at reconstruction by the calculation of an object coded in a hologram. At that time, 6 h of calculation was required for the reconstruction of a field of 512 × 512 pixels with the Minsk-22 computer, the discrete values being obtained from a holographic plate by 64-bit digitization with a scanner. However, it took until the 1990s for array detector-based digital holography to materialize [SCH 94]. In effect, there have been important developments in two sectors of technology: since this period, microtechnological processes have resulted in charge coupled device (CCD) arrays with sufficiently small pixels to fulfill the Shannon condition for the spatial sampling of a hologram; the

Chapter written by Pascal PICART, Michel GROSS and Pierre MARQUET.

computational treatment of images has become accessible largely due to the significant improvement in microprocessor performance, in particular their processing units as well as storage capacities.

The physical principle of digital holography is similar to that of traditional holography. However, the size of the pixels in an image detector (CCD or complementary metal oxyde semiconductor (CMOS)) is clearly greater than that of the grains of a traditional photographic plate (typically 2–3 µm, compared with some 25 nm). These constraints impose to take into account certain parameters (pixel area, number of pixels and pixel pitch) which were more or less clear in an analog holography.

This chapter, as an introduction to advanced methods detailed in other chapters, aims at describing the different aspects related to digital holography: the principle of light diffraction, how to record a digital hologram and color holograms, algorithms to reconstruct digital holograms, an insight into the different holographic configurations, special techniques to demodulate the hologram, the basic principle of digital holographic interferometry and a brief discussion on tomographic phase imaging.

1.1. Digital holograms

A digital hologram is an interferometric mixing between a reference wave and a wave from the object of interest. This section presents the basic properties related to a digital hologram.

1.1.1. *Interferences between the object and reference waves*

Figure 1.1 illustrates the basic geometry for recording a digital hologram. An object wave is coherently mixed with a reference wave, and their interferences are recorded in the recording plane H. In digital holography, the recording is performed by using a pixel matrix sensor.

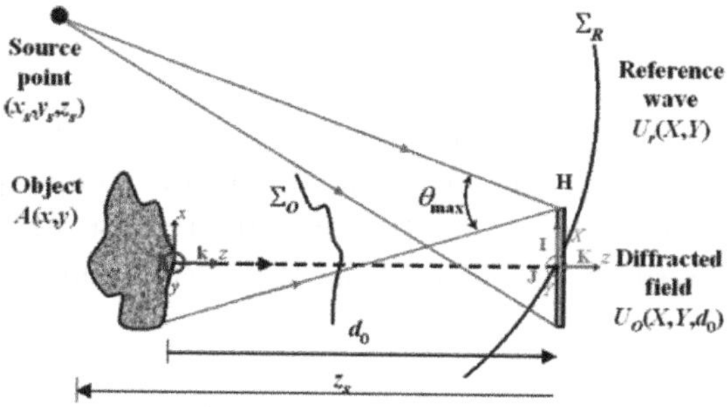

Figure 1.1. *Free space diffraction, interferences and notations. For a color version of this figure, see www.iste.co.uk/picart/digiholography.zip*

Consider an extended object illuminated with a monochromatic wave. This object diffracts a wave to the observation plane localized at a distance $d_0 = |z_0|$. The surface of the object generates a wavefront which will be denoted as A ($i = \sqrt{-1}$):

$$A(x,y) = A_0(x,y)\exp\left(i\psi_0(x,y)\right). \qquad [1.1]$$

The amplitude A_0 describes the reflectivity/transmission of the object and phase ψ_0 is related to its surface and shape or thickness and refractive index. Because of the natural roughness of the object, ψ_0 is a random variable, uniformly distributed over $[-\pi,+\pi]$. The diffracted field U_O at distance d_0, and at spatial coordinates (X,Y) of the observation plane, is given by the propagation of the object wave to the recording plane. In the observation plane, this wave can be simply written as:

$$U_O(X,Y,d_0) = a_O(X,Y)\exp\left(i\phi_O(X,Y)\right), \qquad [1.2]$$

here a_O is the modulus of the complex amplitude and φ_O is its optical phase. Since the object is naturally rough, the diffracted field at distance d_0 is a speckle pattern [DAI 84, GOO 07].

Let us consider U_r, the complex amplitude of the reference wavefront, at the recording plane. We have:

$$U_r(X,Y) = a_r(X,Y)\exp\left(i\phi_r(X,Y)\right), \qquad [1.3]$$

where a_r is the modulus and φ_r is the optical phase. The reference wavefront usually comes from a small pinhole: thus, it is a spherical divergent wave, impacting the plane with a non-zero incident angle. Considering (x_s, y_s, z_s) the coordinates of the source point in the hologram reference frame ($z_s<0$), the optical phase of the reference wave can be written in the paraxial approximations by [GOO 72, GOO 05]:

$$\phi_r(X,Y) \cong \frac{\pi}{\lambda z_s}\left((X-x_s)^2+(Y-y_s)^2\right). \qquad [1.4]$$

This optical phase can also be written as:

$$\phi_r(X,Y) = 2\pi(u_0 X + v_0 Y) + \frac{\pi}{\lambda z_s}(X^2+Y^2) + \phi_s, \qquad [1.5]$$

where (u_0,v_0) are the carrier spatial frequencies of the hologram, and φ_s is a constant term that can be omitted. When $(u_0,v_0) = (0,0)$, i.e. reference point source localized on the z-axis, holography is said to be "in-line" (no tilt between the two waves), whereas when $(u_0,v_0) \neq (0,0)$, holography is said to be "off-axis" (slight tilt between the two waves). As a general rule, we are interested in adjusting the reference wave so that it has uniform amplitude, i.e. $a_r(X,Y) = C^{te}$. The total illumination, denoted H, is then written as [KRE 96, HAR 02, KRE 04]:

$$H = |U_r + U_O|^2 = |U_r|^2 + |U_O|^2 + U_r^* U_O + U_r U_O^*. \qquad [1.6]$$

This equation can also be written as:

$$H = a_r^2 + a_O^2 + 2a_r a_O \cos(\phi_r - \phi_O). \qquad [1.7]$$

Equations [1.6] and [1.7] constitute what is classically called the *digital hologram*. It includes *three orders*: the 0-order is composed of terms $|U_r|^2+|U_O|^2$, the +1 order is the term $U_r^* U_O$ and the −1 order is the term $U_r U^*_O$, also called the twin image. Generally, the +1 order is of interest because it is related to the initial object, whereas the −1 order exhibits some symmetry that is due to the hermitic property of the Fourier operator. Figure 1.2 shows a digital hologram and a zoom on one of its part.

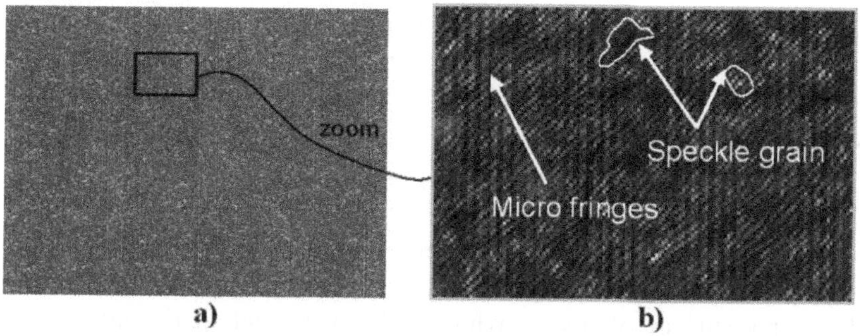

Figure 1.2. *Fine structure of a digital hologram: a) digitally recorded hologram and b) zoom showing micro fringes and speckle grains*

As can be seen, the microstructure of a digital hologram is composed of micro fringes, on the one hand, and light grains, on the other hand. These light grains are *speckles* that are due to the random nature of the light reflected from the object [GOO 85, DAI 84]. Note that in the case where the object is transparent and non-diffusing, the speckle nature of the hologram may disappear.

1.1.2. *Role of the image sensor*

1.1.2.1. *Spatial sampling and Shannon conditions*

In digital holography, the hologram is recorded with $N \times M$ pixels having pitches p_x and p_y and active surfaces Δ_x and Δ_y, respectively, for the x- and y-directions. Thus, the space coordinates in the recording plane are sampled; this means that we have $(X, Y) = (np_x, mp_y)$ with $(m; n) \in (-M/2, +M/2-1; -N/2, +N/2-1)$. In addition, the sampling of the hologram leads to Shannon conditions. Considering the maximum angle $\theta_{\max}$ (see Figure 1.1) between the two waves, the micro fringes locally produced by the two tilted wavefronts must be sampled so that the sampling pitch is at least equal to two times the fringe period. Thus, this leads to the maximum acceptable angle for the setup, according to the following equation [SCH 94]:

$$\theta_{\max} < 2\sin^{-1}\left(\frac{\lambda}{4\max\left(p_x, p_y\right)}\right). \qquad [1.8]$$

For example, with λ = 632.8 nm and $p_x = p_y$ = 4.65 μm, the maximum acceptable angle is smaller than 4°. This means that the setup must be precisely adjusted so as to fulfill the Shannon conditions.

1.1.2.2. *Low-pass filtering*

The digital hologram effectively recorded by the sensor is not simply described by equation [1.6]. Indeed, we must take into account the active surface of pixels, which induces a local spatial integration. So, the recorded hologram at point (np_x, mp_y) was given to be written as [PIC 08]:

$$H_{PIX}\left(np_x, mp_y\right) = \left[H(X,Y) * \Pi_{\Delta_x,\Delta_y}(X,Y)\right]\left(np_x, mp_y\right). \quad [1.9]$$

with the even pixel function:

$$\Pi_{\Delta_x,\Delta_y}(x,y) = \begin{cases} 1/\Delta_x \times 1/\Delta_y & \text{if } |x| \leq \Delta_x/2 \text{ and } |y| \leq \Delta_y/2 \\ 0 & \text{if not} \end{cases}. \quad [1.10]$$

From equation [1.9], the basic effect can be understood: since the pixel provides local integration of the micro fringes, the consequence is a blurring of these fringes. Qualitatively, this means that the spatial resolution will deteriorate and that the pixel induces a low-pass filtering to the digital hologram.

1.1.2.3. *Effect of the exposure time*

During the recording of the hologram, the pixel receives light for a certain duration, called the exposure time T. The total energy received by the sensor is such that [KRE 96]:

$$W = \int_{t_1}^{t_1+T} H(t)\,dt. \quad [1.11]$$

When the hologram has no temporal dependence, the time integration can be omitted. However, in the case where the object exhibits time dependence, i.e. sinusoidal oscillation, the exposure time influences significantly the recorded hologram. The characteristic parameter of the recording is the cyclic ratio defined by $\alpha = T/T_0$, which is the ratio between the exposure time T and the oscillation period T_0. Typically, if $\alpha \ll 1$, the recording regime uses light pulses and is equivalent to a freezing of the object at the instant at which the recording is performed ("impulse regime") [LEC 13]. When, on the contrary, we have $\alpha \gg 1$, the regime is said to be "time-averaging". The object reconstructed from the digital hologram is then amplitude-modulated by a Bessel function [PIC 03, PIC 05]. In experiments for which $0 \ll \alpha < 1$, the cyclic ratio is too high to be classified as "impulse" and too low to be considered as "time-averaging". This intermediary regime is called "quasi-time-averaging" and is completely described in [LEC 13]. The object amplitude also exhibits a modulation that is more complex than that of the pure time-averaging regime.

1.1.2.4. *Recording digital color holograms*

The first digital color holograms appeared in the 2000s with the advent of color detectors. Yamaguchi showed the applicability of digital color holography to the color reconstruction of objects [YAM 02]. Since then, numerous applications have been developed, particularly in the domain of contactless metrology: flow analysis in fluid mechanics [DEM 03, DES 08, DES 12], surface profilometry by two-color microscopy [KUM 09, KUH 07, MAN 08], three-color digital holographic microscopy (DHM) even with low coherence [DUB 12] and multidimensional metrology of deformed objects [KHM 08, TAN 10a, TAN 10b, TAN 11]. There are different approaches for recording digital color holograms, in particular for *simultaneously* recording the three colors. The simplest method consists of using a monochromatic detector and recording the colors *sequentially*. This method was proposed by Demoli in 2003 [DEM 03] and is only adapted to the case of objects which vary slowing in time. Figure 1.3 illustrates the different recording strategies. The first possibility consists of using a chromatic filter organized in a Bayer mosaic (Figure 1.3(a)). However, in such a detector, half of the pixels detect green, and only a quarter detect red or blue [YAM 02, DES 11].

The spatial color filter creates holes in the mesh, and therefore a loss of information, which translates into a loss of resolution. For example, Yamaguchi used a detector with 1,636 × 1,238 pixels of size 3.9 × 3.9 μm^2 [YAM 02], and his results had a relatively low spatial resolution. The number of pixels for each color was 818 × 619, and the pixel pitch was 7.8 µm. The second possibility consists of using three detectors organized as a "tri-CCD", the spectral selection being carried out by a prism with dichroic layers (Figure 1.3(b)). Such a detector guarantees a high spatial resolution and a spectral selectivity compatible with the constraints of digital color holography. Of course, the relative adjustment of the three sensors must be realized with high precision. For example, Desse developed a type of holographic color interferometry for use in fluid mechanics, with three detectors of 1,344 × 1,024 pixels of size 6.45 µm × 6.45 µm [DES 11]. The third possibility consists of using a color detector based on a stack of photodiodes [TAN 10a, TAN 10b, DES 08, DES 11] (http://www.foveon.com, Figure 1.3(c)). The spectral selectivity is relative to the mean penetration depth of the photons in the silicon: blue photons at 425 nm penetrate to around 0.2 µm, green photons at 532 nm to around 2 µm and red photons at 630 nm to around 3 µm. Thus, the construction of junctions at depths at around 0.2, 0.8 and 3.0 µm gives the correct spectral selectivity for color imaging. However, the spectral selectivity is not perfect, as green photons may be detected in the blue and red bands, but the architecture guarantees a maximum spectral resolution since the number of effective pixels for each wavelength is that of the entire matrix. For example, [TAN 10b] uses a stack of photodiodes with 1,060 × 1,414 pixels of size 5 × 5 μm^2. One last possibility consists of using a monochromatic detector combined with spatial chromatic multiplexing (Figure 1.3(d)). Each reference wave must have different separately adjusted spatial frequencies according to their wavelengths. The complexity of the experimental apparatus increases with the number of colors. For two-color digital holography, it is acceptable; for three colors, it becomes prohibitive. A demonstration of this approach is given in [PIC 09, MAN 08, KUH 07] and [TAN 11].

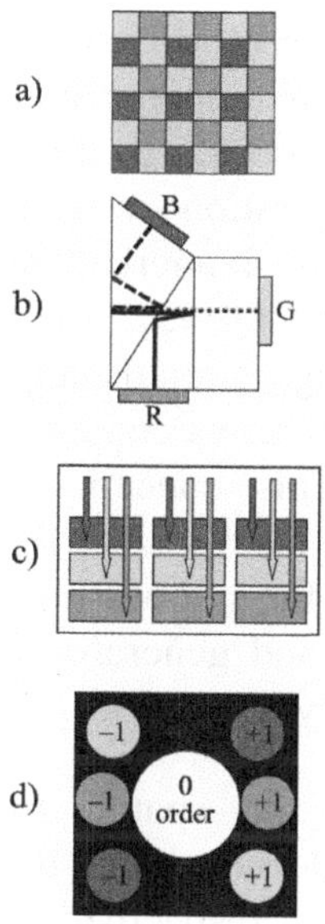

3 color recording with a Bayer mosaïc
⇒ low spatial resolution
⇒ medium spectral selectivity

3 color recording with a tri-CCD
⇒ 3 matrices of pixels
⇒ good spectral selectivity
⇒ need of spatial adjustment

3 color recording with a stack of photodiodes
⇒ spatial resolution is maximal
⇒ need of color calibration

3 color recording with a monochrome CCD
⇒ 3 reference waves
⇒ spectral selectivity by spatial multiplexing

Figure 1.3. *Recording digital color holograms. For a color version of this figure, see www.iste.co.uk/picart/digiholography.zip*

1.1.3. *Demodulation of digital holograms*

Equations [1.6] and [1.7] describe the digital hologram. The +1 order is of interest because it includes the object wave through term $U_r^* U_O$. Note that the −1, $U_r U^*_O$, is the complex conjugate of $U_r^* U_O$ and that it includes also the full information on the object wave. The demodulation of the digital hologram consists of retrieving the +1 order from the recording of H. There are mainly two ways to perform demodulation: using slightly off-axis geometry at the recording, or using phase-shifting [CUC 99b]. These approaches are detailed in the next sections.

1.1.3.1. *Off-axis holograms*

Off-axis geometry introduces a spatial carrier frequency and demodulation restores the full spatial frequency content of the wavefront, i.e. $U_r^* U_O$. In equation [1.5], the phase of the reference wave includes the carrier spatial frequencies of the hologram (u_0,v_0). When $(u_0,v_0) \neq (0,0)$, there is a slight tilt between the two waves and holography is off-axis. Practically, the different diffraction terms encoded in the hologram (zero-order wave, real image and virtual image) are propagating in different directions, enabling their separation for reconstruction. This configuration was the one employed for the first demonstration of a fully numerical recording and reconstruction holography [SCH 94, COQ 95]. In practice, reconstruction methods based on off-axis configuration usually rely on Fourier methods to filter one of the diffraction terms contained in the hologram ($U_r^* U_O$ or $U_r U^*_O$) [CUC 00]. This concept was first proposed by Takeda *et al.* [TAK 82] in the context of interferometric topography. The method was later extended for smooth topographic measurements for phase recovery [KRE 86] and generalized for the use in DHM with amplitude and phase recovery [CUC 99a].

According to equations [1.3]–[1.6], in the spatial frequency spectrum, a three-modal distribution is related to the three diffraction orders of the hologram (FT and FT^{1} means, respectively, Fourier transform and inverse Fourier transform):

$$FT[H](u,v)=C_0(u,v)+C_1(u-u_0,v-v_0)+C_1^*(-u-u_0,-v-v_0), \qquad [1.12]$$

where C_0 is the Fourier transform of the zero-order and C_1 is the Fourier transform of the +1 order. If the three orders are well separated in the Fourier plane, the +1 order can be extracted from the Fourier spectrum. Figure 1.4 illustrates the spectral distribution in the Fourier domain of the digital hologram. The spatial frequencies (u_0,v_0) localize the useful information and they must be adjusted to minimize the overlapping of the three diffraction orders. By applying a bandwidth-limited filter ($\Delta u \times \Delta v$ width) around the spatial frequency (u_0,v_0), and after filtering and inverse two-dimensional (2D) Fourier transform, we get the object complex amplitude:

$$\begin{aligned} O_{+1}(x,y) &= FT^{-1}\left[C_1(u-u_0,v-v_0)\right] \\ &\cong a_r\left\{a_O(x,y)\exp\left[i\phi_O(x,y)\right]\exp\left[2i\pi(u_0x+v_0y)\right]\right\}*h(x,y) \end{aligned}, \qquad [1.13]$$

where the symbol * means convolution and $h(x,y)$ is the impulse response corresponding to the filtering applied in the Fourier domain.

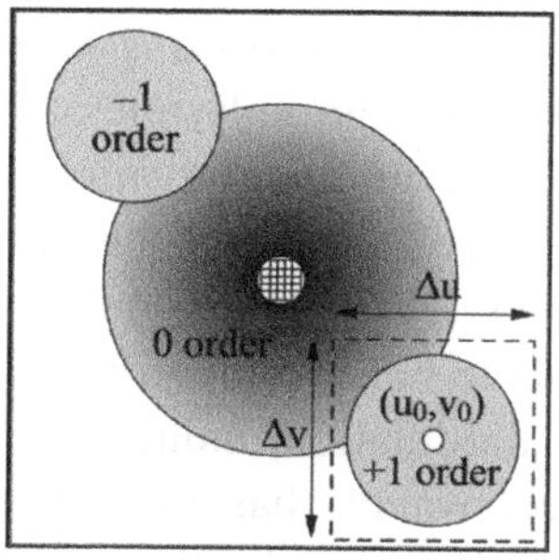

Figure 1.4. *Spectral distribution of orders and spectral filtering*

The impulse response of the filter is such that:

$$h(x,y)=\Delta u\Delta v\exp\left[2i\pi\left(u_0x+v_0y\right)\right]\operatorname{sinc}\left(\pi\Delta ux\right)\operatorname{sinc}\left(\pi\Delta vy\right), \quad [1.14]$$

The spatial resolution is then related to $1/\Delta u$ and $1/\Delta v$, respectively, in the x-y axis. In addition, the phase recovered with equation [1.13] includes the spatial carrier modulation that has to be removed. This may be achieved by multiplying O_{+1} by $\exp[-2i\pi(u_0x+v_0y)]$.

Note that a filter having a circular bandwidth (instead of a rectangular bandwidth) can also be used [CUC 99a]. In that case, the impulse response of the filter is proportional to a J_0 Bessel function.

Then, the optical object phase at the hologram plane can be estimated from relation:

$$\phi_O(x,y)=\tan^{-1}\left(\frac{\Im_m\left[O_{+1}(x,y)\right]}{\Re_e\left[O_{+1}(x,y)\right]}\right), \quad [1.15]$$

and the object amplitude by:

$$a_O(x,y)=\sqrt{\Im_m^2\left[O_{+1}(x,y)\right]+\Re_e^2\left[O_{+1}(x,y)\right]}, \quad [1.16]$$

In equations [1.15] and [1.16], $\Im_m[...]$ and $\Re_e[...]$, respectively, mean the imaginary and real parts of the complex value.

The main advantage of this approach is its capability of recovering the complex object wave through only one acquisition. Thus, there is no time spent heterodyning or moving mirrors and the influence of vibrations is greatly reduced. However, as the diffraction terms are spatially encoded in the hologram, this one shot capability potentially comes at the cost of usable bandwidth (filter with width $\Delta u \times \Delta v$). In addition, the frequency modulation, induced by the angle between the reference and the object waves, has to guarantee the separation of the information contained in the different diffraction terms that are encoded in the hologram while carrying a frequency compatible with the sampling capacity of digital detectors.

However, in the field of microscopy, the microscope objective usually allows us to properly adapt the object wave field to the sampling capacity of the camera. Definitively, the lateral components of the wave vector $k_{x \text{ or } y}$ are divided by the magnification factor M of the microscope objective. Practically, when a standard camera with pixels at a few microns is used, microscope objective with magnification larger than ×20 makes possible obtaining diffraction-limited resolution even when high numerical apertures (NAs) are considered [MAR 05]. It should also be mentioned that the numerical reconstruction of the object wavefront, particularly its propagation, represents a breakthrough in modern optics and specifically in microscopy [MAR 13]. Indeed, in addition to the possibility to achieve off-line autofocusing [LAN 08, LIE 04a, LIE 04c, DUB 06a] and to extend the depth of focus [FER 05], these numerical reconstruction procedures permit us to mimic complex optical systems as well as to compensate for aberrations [COL 06a, COL 06b], distortions and experimental noise leading to the development of various simplified and robust interferometric configurations able to quantitatively measure optical path lengths with ultrahigh resolution [MAR 13, LEE 13], in practice down to the subnanometer scale [KUH 08], depending on the wavelength and other parameters including the integration time.

1.1.3.2. *Phase-shifting digital holography*

In contrast to off-axis digital holography (Fourier domain), the complex amplitude of the object wave can be directly extracted by using phase-shifting methods in the temporal domain [CRE 88, DOR 99]. This approach was described by Yamaguchi in 1997 [YAM 97, YAM 01a, YAM 01b] and leads to the reconstruction of an image free from the zero-order and of the twin image (−1 order). Consider the hologram equation written as:

$$H = a_r^2 + a_O^2 + 2a_r a_O \cos(\phi_O - \phi_r) \qquad [1.17]$$

Basically, in equation [1.17] we should consider three unknowns: the offset term $a_r^2+a_O^2$, the modulation term $2a_r a_O$ and the phase of the cosine function, $\varphi_O-\varphi_r$. So, with at least three values for H, we should be able to solve these three unknowns. This can be done by shifting the phase in the cosine function, by adding in the holographic interferometer a phase modulator. Practically, a piezoelectric transducer (PZT) is used (although other methods do exist) [CRE 88, DOR 99]. The PZT is stamped to a mirror and applying a small voltage to the PZT has to slightly move the mirror as a consequence, and thus to shift the optical phase. With at least three positions of the mirror, the object wave field can be recovered. The robustness of the method increases with increasing the number of phase-shifted holograms. Consider a phase-shifted hologram with a phase-shift being an integer division of 2π, i.e. $2\pi/P$, with P an integer. We have:

$$H_p = a_r^2 + a_O^2 + 2a_r a_O \cos\left(\phi_O - \phi_r + 2(p-1)\pi / P\right) \qquad [1.18]$$

with $p = 1, 2, \ldots, P$. For $P \geq 3$, the phase of the object wave in the detector plane may be calculated by [GRE 84]:

$$\phi_O = \phi_r + \arctan\left\{\frac{\sum_{n=1}^{P} H_p \sin\left(2\pi(p-1)/P\right)}{\sum_{n=1}^{P} H_p \cos\left(2\pi(p-1)/P\right)}\right\} \qquad [1.19]$$

and the amplitude is calculated by:

$$a_O = \frac{1}{2a_r}\sqrt{\left(\sum_{p=1}^{P} H_p \sin\left(2\pi(p-1)/P\right)\right)^2 + \left(\sum_{p=1}^{P} H_p \cos\left(2\pi(p-1)/P\right)\right)^2} \qquad [1.20]$$

If the reference wave is plane or spherical, that is free from aberrations, the phase $\phi_O(x,y)$ may be determined without ambiguity and compensated. The complex wave may be evaluated and the object may be directly reconstructed. Using the conjugate complex wave, we may calculate the twin image.

With $P = 4$, we obtain the most widely used method, which was proposed by Yamaguchi and Zhang [YAM 97], using four π/2 phase-shifted holograms [CRE 88, WYA 75]. In this case, we have:

$$\phi_O = \phi_r + \arctan\left\{\frac{H_4 - H_2}{H_1 - H_3}\right\}$$

$$a_O = \frac{1}{4a_r}\sqrt{[H_1 - H_3]^2 + [H_4 - H_2]^2} \qquad [1.21]$$

1.1.3.3. *Parallel phase-shifting*

In the technique of phase-shifting, both the single-shot and real-time capability of digital holography are lost because of the sequential recording of holograms. The four holograms are sequentially recorded by using reference waves with different phase retardations, such as 0, π/2, π and 3π/2. Although the phase-shifting method achieves noiseless images, it is useless for the instantaneous measurement of moving objects. Even though off-axis digital holography is one candidate for instantaneously obtaining only the first-order diffracted wave, it has some drawbacks: a high-resolution image sensor is required to record spatial carrier fringes and the spatial bandwidth has to be judiciously occupied (see Figure 1.4). In parallel phase-shifting digital holography, the four kinds of phase-shifting are simultaneously carried out for the reference wave in each segment consisting of 2 × 2 pixels of the image sensor in the recording hologram; thus, it implements four phase-shifting processes by using a spatial division-multiplexing technique. The four holograms required for the phase-shifting interferometry are numerically generated from a hologram recorded with the reference wave. The recording process of the technique is schematically illustrated in Figure 1.5 [AWA 06a, AWA 6b].

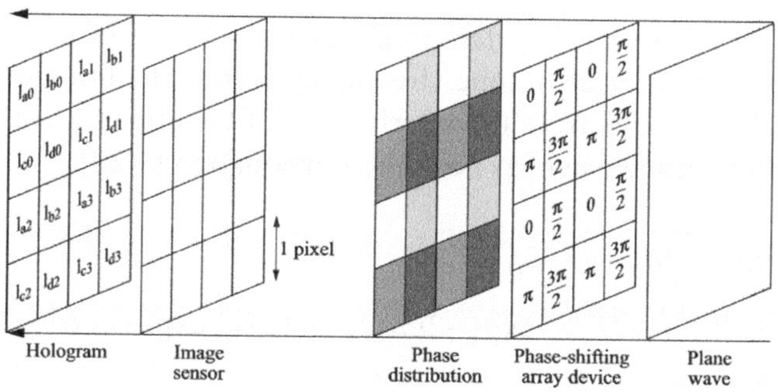

Figure 1.5. *Implementation of parallel phase-shifting digital holography, phase-shifting array device and the distribution of the reference wave for parallel four-step phase-shifting (from [AWA 06a])*

A phase-shifting device array is placed in the reference beam in the holographic interferometer. The array device is a segmented array with a 2 × 2 cell configuration that generates the periodic four-step phase distributions 0, $\pi/2$, π and $3\pi/2$. The array device can be implemented by using a glass plate with a periodic four-step thickness. The array device is imaged onto the image sensor so that the phase distribution of the reference wave at the image sensor plane corresponds with the arrangement of pixels in the image sensor. The size of the imaged cells at the image sensor is the same as that of the pixels. Thus, the image sensor captures a hologram recorded with the reference wave containing the four-step phase distributions. The pixels containing the same phase-shift are extracted from the recorded hologram. For each phase-shift, the extracted pixels are relocated in another 2D image at the same addresses at which they were located before being extracted. The values of the pixels not relocated in the 2D image are simply linearly interpolated by using the adjacent pixel values in the reconstruction process. By carrying out this relocation and interpolation for the four phase-shifts, four holograms H_1, H_2, H_3 and H_4 are obtained. Then, the amplitude and phase of the complex object field can be calculated using the conventional algorithm [1.21].

1.1.3.4. *Heterodyne digital holography*

In a heterodyne digital holographic scheme, the reference beam is dynamically phase-shifted with respect to the object field. This shift produces time-varying interferograms at the sensor plane. Generally, the phase-shift is linear in time (frequency shift). The hologram in the detector plane results from the interference of the object wave with the δf-shifted reference wave, as described in equation [1.22]:

$$\begin{aligned} H &= \left|U_r \exp(2i\pi\delta f t) + U_O\right|^2 \\ &= \left|U_r\right|^2 + \left|U_O\right|^2 + U_r^* U_O \exp(-2i\pi\delta f t) + U_r U_O^* \exp(+2i\pi\delta f t) \end{aligned} . \quad [1.22]$$

A set of P holograms H_p ($p = 1, 2, \ldots, P$) is recorded within a δf period at $t_p = 2\pi(p-1)/\delta f$ [LE 00, LE 01]. The demodulation algorithm is then:

$$U_O = \frac{1}{P}\frac{1}{U_r^*}\sum_{p=1}^{p=P} H_p \exp\left[\frac{2i\pi(p-1)}{P}\right]. \quad [1.23]$$

For $P = 4$, the object complex wave is proportional to $(H_1-H_3)+i(H_2-H_4)$ and the algorithm is quite similar to that provided in equation [1.21]. Thus, heterodyne holography measures the phase, using the information obtained at different times. From the practical point of view, the frequency shift is provided by combining two acousto-optic modulators working at $\Delta f+\delta f$ and Δf, respectively, and δf is adjusted to be equal to one quarter of the sensor frame rate [LE 00, LE 01, ABS 10, SAM 11].

Combining off-axis holography with heterodyning permits us to reach the shot-noise detection and to achieve the ultimate sensitivity of digital holography [ATL 07, ATL 08, GRO 07, GRO 08].

1.2. Back-propagation to the object plane

The previous sections have discussed on the basics of digital hologram recording and demodulation. In order to discuss the digital reconstruction of the object wave at the object plane (and not necessarily at the sensor plane), this section presents the basics of the scalar diffraction of light. Algorithms used to back-propagate the object field estimated at the sensor plane are based on this approach.

1.2.1. *Monochromatic spherical and plane waves*

In this chapter, the time dependence of the optical wavefront is implicit and not included in the equations describing the diffraction. In a space described by a Cartesian coordinate system *Oxyz*, a point *P* is described by a set of three coordinates (x, y, z) and we will use the modulus of the distance **OP** by $r=|\mathbf{OP}|=\sqrt{x^2+y^2+z^2}$. We also use $k = 2\pi/\lambda$, the modulus of the wave vector, here λ is the wavelength of the light. We assume that the optical field is written as:

$$U(P,t)=U(P)\exp(-2i\pi\nu t) \qquad [1.24]$$

where $U(P)$ is the complex amplitude at the observation point $P(x, y, z)$ and ν is the frequency of the light wave. Let us begin with the definition of a spherical wave. If the point source of a spherical wave is at the origin of a Cartesian coordinate system, the complex amplitude of a spherical wave can, therefore, be expressed by [GOO 72, COL 70, YAR 85]:

$$U(x,y,z)=\begin{cases}\dfrac{A_0}{r}\exp(ikr) & (divergent)\\[2mm] \dfrac{A_0}{r}\exp(-ikr) & (convergent)\end{cases} \qquad [1.25]$$

We note that the amplitude is proportional to the inverse of the distance between the point source and r, the observation point. When the center of the spherical wave is at point (x_c, y_c, z_c), instead of the origin, the expressions are identical, with r substituted for:

$$r=\sqrt{(x-x_c)^2+(y-y_c)^2+(z-z_c)^2} \qquad [1.26]$$

For a plane wave propagating in a homogeneous medium, the wavefront is perpendicular to the propagation direction. The plane wave can be written as:

$$U(x,y,z)=A_0\exp\left[ik(x\cos\alpha+y\cos\beta+z\cos\gamma)\right] \qquad [1.27]$$

The propagation direction is defined by the direction cosines $\cos\alpha$, $\cos\beta$ and $\cos\gamma$ of [1.27]. This relation shows that for a real number C, the expression $x\cos\alpha + y\cos\beta + z\cos\gamma = C$ describes a phase plane whose normal is in the direction given by the cosines $\cos\alpha$, $\cos\beta$ and $\cos\gamma$. Since different values of C correspond to different parallel planes, expression [1.27] represents a wave propagating in the direction normal to these planes.

Figure 1.6 illustrates the concept of spherical and plane waves. Figure 1.6(a) shows a spherical wavefront with center A, that emits a divergent spherical wavefront Σ (see [1.25]). In a homogeneous medium, rays are perpendicular to Σ, and the wave is deformed when propagating to the right (to the left for a convergent wavefront). When the point source tends to infinity, the spherical wave tends to a plane wave, as illustrated in Figure 1.6(b). In this case, the rays become parallel and the beam propagates without any deformation.

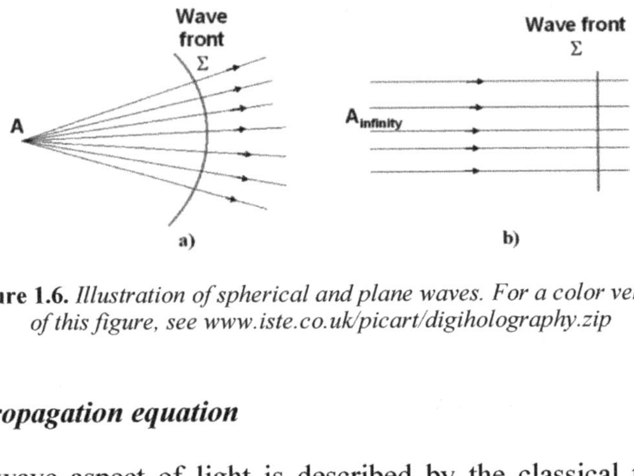

Figure 1.6. *Illustration of spherical and plane waves. For a color version of this figure, see www.iste.co.uk/picart/digiholography.zip*

1.2.2. *Propagation equation*

The wave aspect of light is described by the classical theory of electromagnetism, by the Maxwell's equations [BOR 99, GOO 72, LAU 10, YAR 85]. In this chapter, we consider the case of a homogeneous medium. After some mathematics, Maxwell's equations can be reduced to this propagation equation:

$$\nabla^2 \boldsymbol{E} - \frac{1}{c^2}\frac{\partial^2 \boldsymbol{E}}{\partial t^2} = 0, \qquad [1.28]$$

where **E** is the electric field and c is the velocity of light in the medium. Operator $\nabla^2 = \partial^2 / \partial x^2 + \partial^2 / \partial y^2 + \partial^2 / \partial z^2$ is the Laplacian operator. Note that [1.28] is also valid for the magnetic field **B**.

1.2.3. *Angular spectrum transfer function*

Substituting [1.24] into [1.28], we obtain an equation which is independent of time t, known as the Helmholtz equation:

$$\left(\nabla^2 + k^2\right)U(P) = 0 \qquad [1.29]$$

This equation can be solved in the Fourier domain. We suppose that z is the distance between the initial and observation planes, and that $U(x,y,0)$ and $U(x, y, z)$ are the respective complex amplitudes of these two planes. Moreover, in the frequency space, their spectral functions are $G_0(u,v)$ and $G_z(u,v)$, respectively, (u,v) being the spatial frequencies associated with the spatial coordinates (x, y). These two functions are defined by:

$$G_0(u,v) = \int_{-\infty}^{\infty}\int_{-\infty}^{\infty} U(x,y,0)\exp\left[-2i\pi(ux+vy)\right]dxdy \qquad [1.30]$$

$$G_z(u,v) = \int_{-\infty}^{\infty}\int_{-\infty}^{\infty} U(x,y,z)\exp\left[-2i\pi(ux+vy)\right]dxdy \qquad [1.31]$$

The demonstration will not be provided in this chapter; a general solution to the differential equation can be expressed with the Fourier components of $U(x,y,0)$ and $U(x, y, z)$ according to:

$$G_z(u,v) = G_0(u,v)\exp\left[\frac{2i\pi}{\lambda}z\sqrt{1-(\lambda u)^2-(\lambda v)^2}\right] \qquad [1.32]$$

Then the complex field at distance z can be obtained by:

$$U(x,y,z) = \int_{-\infty}^{\infty}\int_{-\infty}^{\infty} G_z(u,v)\exp\left[2i\pi(ux+vy)\right]dudv \qquad [1.33]$$

So, we have a relation between the spectrum of the wave in the initial plane and that obtain in the observation plane. This relation shows that, in the frequency space, the spectral variation in complex amplitude caused by the propagation of light over the distance z is represented by its multiplication by a phase-delay factor:

$$G(u,v) = \exp\left[2i\pi z / \lambda\sqrt{1-(\lambda u)^2-(\lambda v)^2}\right] \quad [1.34]$$

According to the theory of linear systems, the process of diffraction is a transformation of the light field across an optical system, as the phase-delay factor can be interpreted as a transfer function in the frequency space. This interpretation of the propagation of light is called *the propagation of the angular spectrum* and the associated transfer function [1.34] is called *the angular spectrum transfer function*. Figure 1.7 illustrates this approach.

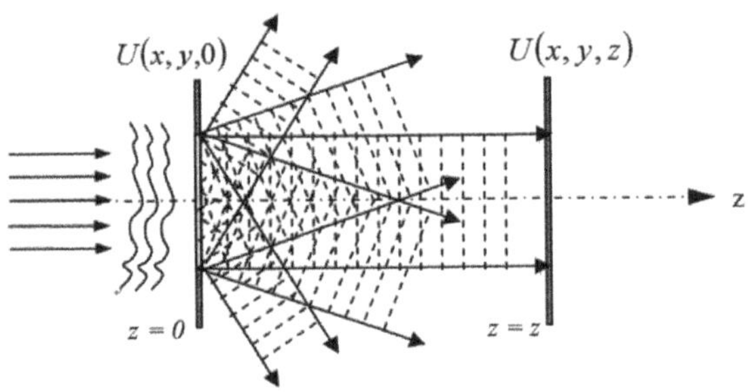

Figure 1.7. *Scheme of the diffraction by the angular spectrum*

Figure 1.7 means that that the field $U(x,y,z)$ can be considered as a superposition of plane waves of amplitude $G_z(u,v)dudv$ propagating in a direction whose cosines are $\{\cos\alpha, \cos\beta, \cos\gamma\} = \left\{\lambda u, \lambda v, \sqrt{1-(\lambda u)^2-(\lambda v)^2}\right\}$. From the diffraction of the angular spectrum, [1.34] means that the elementary waves satisfying $1-(\lambda u)^2-(\lambda v)^2<0$ are attenuated by the propagation, i.e. all the components satisfying this relation only exist in a zone very close to the initial plane. These components of the angular spectrum are, therefore, called "evanescent waves". As the components of the observation plane must satisfy the relation $1-(\lambda u)^2-(\lambda v)^2>0$, i.e. $u^2+v^2<1/\lambda^2$, propagation in free space can be considered as an ideal low-pass filter of radius $1/\lambda$ in the frequency space. Consequently, on the condition that we can obtain the spectrum of $U(x,y,0)$, the spectrum in the observation plane, $U(x, y, z)$ can be expressed by relation [1.32]. Using the direct and inverse Fourier transforms (FT and FT^{1}), the diffraction calculation process can be described as:

$$U(x,y,z)=FT^{-1}\left\{FT\{U(x,y,0)\}\exp\left[\frac{2i\pi}{\lambda}z\sqrt{1-(\lambda u)^2-(\lambda v)^2}\right]\right\} \qquad [1.35]$$

1.2.4. *Kirchhoff and Rayleigh–Sommerfeld formulas*

There also exist two more solutions to the Helmholtz equation: Kirchhoff's formula and that of Rayleigh–Sommerfeld. Using the coordinates shown in Figure 1.8 which represents the relationship between the initial plane and the observation plane, these two formulas are written in the same mathematical expression [GOO 05]:

$$U(x,y,d_0)=\frac{1}{i\lambda}\int_{-\infty}^{\infty}\int_{-\infty}^{\infty}U(X,Y,0)\frac{\exp(ikr)}{r}K(\theta)dXdY \qquad [1.36]$$

where

$$r=\sqrt{(x-X)^2+(y-Y)^2+d_0^2} \qquad [1.37]$$

and θ is the angle between the normal at point $(X,Y,0)$, and the vector **MP** from point $(X,Y,0)$ to point (x, y, d_0) (see Figure 1.8), $K(\theta)$ is called the obliquity factor and its three different expressions correspond to three different formulations [GOO 05].

- $K(\theta)=\frac{\cos\theta+1}{2}$ Kirchhoff's formula;
- $K(\theta)=\cos\theta$ first Rayleigh–Sommerfeld solution;
- $K(\theta)=1$ second Rayleigh–Sommerfeld solution.

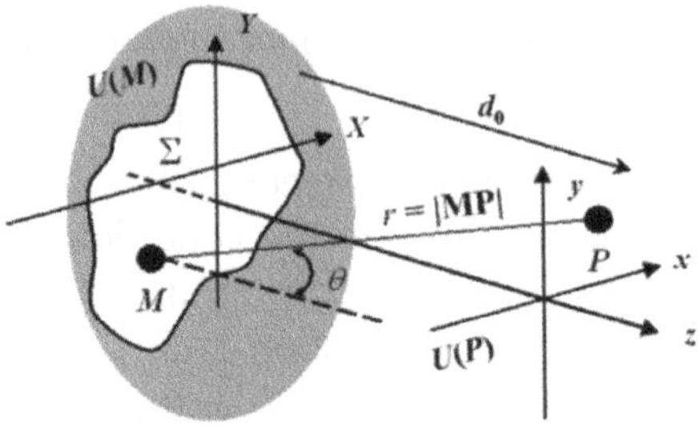

Figure 1.8. *Relation between the initial diffraction plane and the observation plane. For a color version of this figure, see www.iste.co.uk/picart/digiholography.zip*

Even though there exist certain inconsistencies [GOO 05, BOR 99], Kirchhoff's formula gives results in remarkable agreement with experiment, and it is for this reason that it is widely applied in practice. Furthermore, since the angle θ is often small in experimental configurations, the obliquity factors of the three formulations are roughly equal to unity. Thus, the Kirchhoff, Rayleigh–Sommerfeld and angular spectrum formulas are considered as equivalent representations of diffraction. The derivations of the Kirchhoff and Rayleigh–Sommerfeld approaches are presented in detail in [GOO 05]. Readers who would like to go into these aspects in greater detail are invited to familiarize themselves with these appraoches.

1.2.5. *Fresnel approximation and Fresnel diffraction integral*

The equations proposed previously are complex and this is due to the presence of a square root in the complex exponentials. In practice, problems of diffraction quite often concern paraxial propagation, and to simplify the theoretical analysis, we generally use Fresnel's approximation. Let d_0 be the diffraction distance, and expanding the square root in [1.34] to the first order leads to:

$$G(u,v) \cong \exp\left[ikd_0\right]\exp\left[-i\pi\lambda d_0\left(u^2+v^2\right)\right] \qquad [1.38]$$

Given that expression [1.35] can be written in the form of a convolution (* means convolution):

$$U(x,y,d_0) = U(x,y,0) * FT^{-1}\left\{G(u,v)\right\} \qquad [1.39]$$

Substituting [1.38] into [1.39] and knowing that the inverse Fourier transform of $G(u,v)$ is an analytic function, we have:

$$U(x,y,d_0)=U(x,y,0)*\frac{\exp(ikd_0)}{i\lambda d_0}\exp\left[\frac{i\pi}{\lambda d_0}(x^2+y^2)\right], \quad [1.40]$$

In [1.40], we recognize a convolution of $U(x,y,0)$ with the impulse response of free space propagation that will be denoted as $h(x, y, d_0)$:

$$h(x,y,d_0)=-\frac{i}{\lambda d_0}\exp\left[ikd_0\right]\exp\left[\frac{i\pi}{\lambda d_0}(x^2+y^2)\right]. \quad [1.41]$$

Equation [1.41] can also be written as:

$$U(x,y,d_0)=\frac{\exp(ikd_0)}{i\lambda d_0}\int_{-\infty}^{\infty}\int_{-\infty}^{\infty}U(X,Y,0)\exp\left\{\frac{i\pi}{\lambda d_0}\left[(x-X)^2+(y-Y)^2\right]\right\}dXdY \quad [1.42]$$

Equation [1.42] constitutes *Fresnel's diffraction integral.* Note that this approximation consists of replacing spherical wavelets (see [1.25]) by quadratic waves (parabolic surface approximation). Developing the quadratic terms in the exponential of [1.42] leads us to:

$$U(x,y,d_0)=\frac{\exp(ikd_0)}{i\lambda d_0}\exp\left[\frac{i\pi}{\lambda d_0}(x^2+y^2)\right]\times$$
$$\int_{-\infty}^{\infty}\int_{-\infty}^{\infty}U(X,Y,0)\exp\left[\frac{i\pi}{\lambda d_0}(X^2+Y^2)\right]\exp\left[-2i\pi\left(\frac{x}{\lambda d_0}X+\frac{y}{\lambda d_0}Y\right)\right]dXdY \quad [1.43]$$

Thus, with the exception of multiplicative phase and amplitude factors which are independent of X and Y, we can calculate the function $U(x, y, d_0)$ by carrying out the Fourier transform of:

$$U(X,Y,0)\exp\left[\frac{i\pi}{\lambda d_0}(X^2+Y^2)\right] \quad [1.44]$$

This transformation must be evaluated at the frequencies $(u,v) = (x/\lambda d_0, y/\lambda d_0)$ to guarantee a correct spatial scale in the observation plane. The calculation of the two Fresnel diffraction integrals is relatively simple compared to the formulas which rigorously satisfy the Helmholtz equation. In the regime of paraxial propagation, this approximation is relatively precise. By defining the Fresnel transfer function [GOO 05] as:

$$G_F(u,v) = \exp\left[\frac{2i\pi z}{\lambda}\right]\exp\left[-i\lambda z\left(u^2+v^2\right)\right] \quad [1.45]$$

The Fresnel approximation can be expressed by:

$$U(x,y,z) = FT^{-1}\left\{FT\left\{U(x,y,0)\right\}G_F(u,v)\right\} \quad [1.46]$$

This expression is analogous to the angular spectrum formulation [1.35], but the difference is related to the different transfer functions of the two formulas.

The next section discusses the use of the theoretical basics of wave propagation to numerically reconstruct the object wave at the object plane (which is not necessarily the same as the recording plane).

1.3. Numerical reconstruction of digital holograms

1.3.1. *Discrete Fresnel transform*

1.3.1.1. *Algorithm*

The numerical reconstruction with the discrete Fresnel transform is based on the discrete version of equation [1.43] and considering the input plane as the hologram plane. The spatial sampling in the hologram plane $(X,Y) = (np_x, mp_y)$ where $(m; n) \in (-M/2,+M/2-1;-N/2,+N/2-1)$ has to be taken into account. At any distance d_r from the recording plane, the reconstructed object field can be calculated according to [1.47], in which U_O is the object wave at the sensor plane estimated from the demodulation (see section 1.1.3):

$$\begin{aligned} A_r(x,y,d_r) = &-\frac{i}{\lambda d_r}\exp\left(\frac{2i\pi d_r}{\lambda}\right)\exp\left[\frac{i\pi}{\lambda d_r}\left(x^2+y^2\right)\right] \\ &\times\sum_{l=-L/2}^{l=+L/2}\sum_{k=-K/2}^{k=+K/2} U_O\left(lp_x,kp_y\right)\exp\left[\frac{i\pi}{\lambda d_r}\left(l^2p_x^2+k^2p_y^2\right)\right]\times\exp\left[-\frac{2i\pi}{\lambda d_r}\left(lp_x x+kp_y y\right)\right] \end{aligned} \quad [1.47]$$

Note that in off-axis holography, the different diffraction terms encoded in the hologram (zero-order wave, real image and virtual image) are propagated in different directions, thus enabling their separation for reconstruction. This means that equation [1.47] can be directly used with an off-axis hologram (replace U_O by H in [1.47]) to calculate the propagated field at distance d_r. In this case, the reconstructed field appears as illustrated in Figure 1.9.

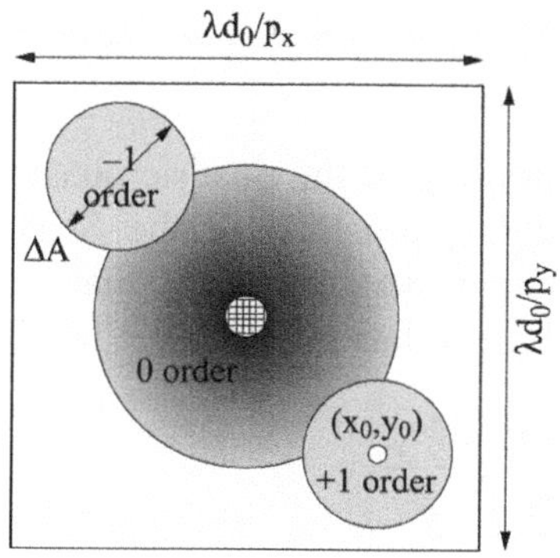

Figure 1.9. *Structure of the reconstructed field of view calculated from an off-axis hologram by using the discrete Fresnel transform*

The spatial localization is related to the spatial carrier frequencies of the hologram, $(x_0, y_0) = (\lambda d_r u_0, \lambda d_r v_0)$, and if the carrier frequencies are correctly adjusted, there is no superposition between the diffractions orders. Then, the object zone can be selected without doubt and the amplitude and phase of the reconstructed object can be estimated.

Since the processor cannot calculate indefinitely, we also have to take into account the spatial sampling in the reconstructed plane. In addition, we can consider that the reconstructed field will be sampled with $(K, L)\geq(M, N)$ pixels. It follows that the sampling pitch is equal to $\Delta\eta = \lambda d_r/Lp_x$ and $\Delta\xi = \lambda d_r/Kp_y$ [KRE 97, YAM 01a]. The spatial sampling in the image plane is simply $x = l\Delta\eta$ and $y = k\Delta\xi$ with l and k varying from $-L/2$ to $L/2-1$ and from $-K/2$ to $K/2-1$. The schematic diagram of the algorithm is given in Figure 1.10 for a reconstruction distance d_r. This algorithm is known as the single-fast Fourier transform (S-FFT) algorithm since it uses only a single FFT computation.

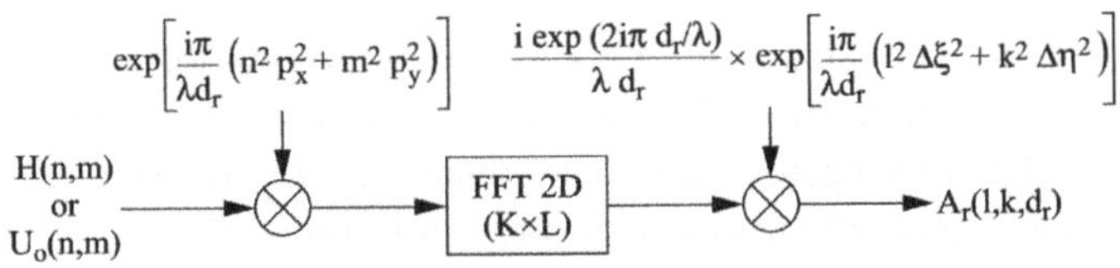

Figure 1.10. *Diagram of the reconstruction with the discrete Fresnel transform*

In addition, the sampling of the quadric phase that is multiplied by the input data (H or U_O) must fulfill the Shannon condition. This means that the minimal distance $d_r^{\min}$ that can be put into the algorithm must fulfill this relation [MAS 03, MAS 99, LI 07]:

$$d_r^{\min} \geq \max\left\{\frac{Np_x^2}{\lambda}, \frac{Mp_y^2}{\lambda}\right\}. \qquad [1.48]$$

Thus, the discrete Fresnel transform cannot be calculated for distance shorter than $d_r^{\min}$.

1.3.1.2. *Spatial resolution in the reconstructed plane*

The computation of the reconstructed field using a finite number of pixels induces a truncate effect. Mathematically, we have to consider the filtering function of the 2D discrete Fourier transform which limits the achievable spatial resolution in the reconstructed plane. It is given by [PIC 08]:

$$\begin{aligned}\tilde{W}_{NM}(x,y,d_r) &= \exp\left[i\pi(N-1)\frac{xp_x}{\lambda d_r} + i\pi(M-1)\frac{yp_y}{\lambda d_r}\right] \\ &\times \frac{\sin(\pi Nxp_x/\lambda d_r)}{\sin(\pi xp_x/\lambda d_r)}\frac{\sin(\pi Myp_y/\lambda d_r)}{\sin(\pi yp_y/\lambda d_r)}.\end{aligned} \qquad [1.49]$$

This function is periodic and its period can be assimilated to a sinc function [PIC 08]. It is interpreted to be the numerical diffraction pattern of the rectangular aperture constituted by the recorded hologram. The width of this function gives the *intrinsic spatial resolutions* in the reconstructed plane. They are given by $\rho_x = \lambda d_r/Np_x$ and $\rho_y = \lambda d_r/Mp_y$, respectively, or the x- and y-directions, and they depend on the wavelength, the number of useful sampling pixels and the reconstruction distance.

1.3.1.3. *Effect of defocus and depth of focus*

Although digital holography is not a conventional imaging method, it exhibits some similarities with classical imaging. Especially, the reconstructed images include a depth of focus. The perfect focus is obtained if the spatial resolution reaches its theoretical limits. The contribution to the degradation of the spatial resolution will not be discussed in this section and the readers may have a look at [YAM 01a, PIC 08, PIC 12]. However, to determine the focal depth of the reconstructed image, we can set the width of the defocusing function as having to be approximately equal to ρ_x. If the perfect image distance is d_i, noting $\Delta z = |d_r - d_i|$, the full depth of focus on both sides of the perfect image plane is given by:

$$2\Delta z \cong \frac{2\lambda d_i^2}{N^2 p_x^2} \quad [1.50]$$

Thus, the focal depth in digital holography is proportional to the square of the angular aperture of the sensor as seen from the object [YAM 01a].

As an illustration, a 2€ coin, 25 mm in diameter, is illuminated at $\lambda = 0.6328$ nm and placed at $d_0 = 660$ mm from the sensor ($N = 1{,}024$, $p_x = 4.65$ μm). The theoretical ideal focused image is obtained with $d_i = -660$ mm and the spatial resolution is $\rho_x = 87.7$ μm in the reconstructed image plane. Then, the depth of focus is estimated at $\Delta z \approx 12.15$ mm. Figure 1.11 provides a set of reconstructed images. Figure 1.11(a) corresponds to the best focus image. Figure 1.11(b) corresponds to an out of focus image with $d_r = d_i - 50$ mm. The image is highly blurred. Figures 1.11(c) and (d) correspond to reconstructions symmetrically included in the depth of focus. As can be seen, the images are focused even if there is a slight difference between the two images. Figures 1.11(e) and (f) correspond to reconstructions slightly outside the depth of focus. The images become to be blurred by the defocusing [PIC 08], whose width increases with the increase of $|d_r – d_i|$. Figure 1.11 shows that if the reconstruction distance is slightly different from the ideal one, there is no significant difference between reconstructed images.

a) best focus : $d_r – d_i = 0$

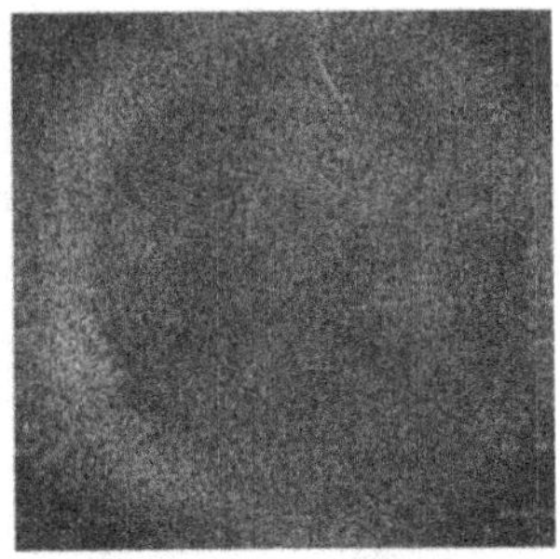

b) $d_r – d_i = –50\text{mm}$

c) $d_r – d_i = –5\text{mm}$

d) $d_r – d_i = +5\text{mm}$

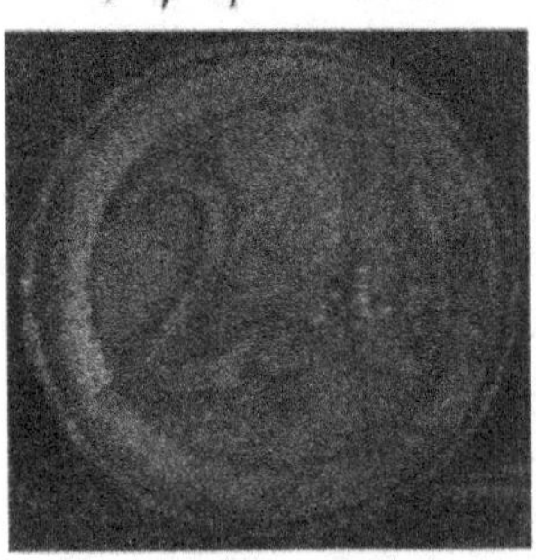

e) $d_r – d_i = –10\text{mm}$

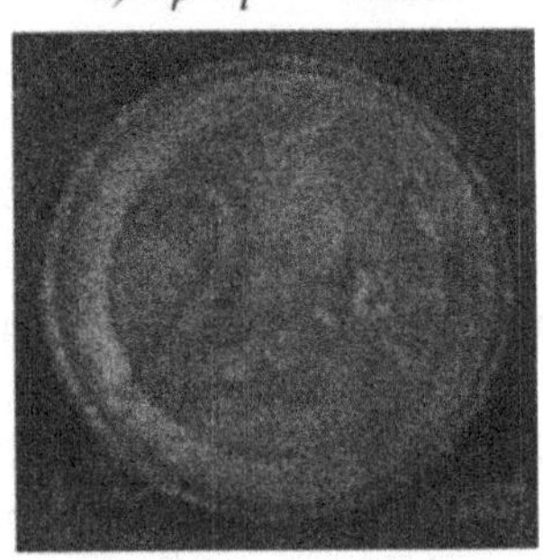

f) $d_r – d_i = +10\text{mm}$

Figure 1.11. *Reconstructed images in and out the depth of focus*

1.3.1.4. *Effect of zero-padding*

Computation of the discrete Fresnel transform with $(K, L) = (N, M)$ uses the data from the raw hologram (H or U_O). If $(K, L) \geq (N, M)$, this case is called "zero-padding" and it consists of adding $(K-N, L-M)$ zeros to the hologram matrix. Fundamentally, these zeros do not add pertinent information; however, they modify the sampling pitches $\Delta\eta = \lambda d_r/Lp_x$ and $\Delta\xi = \lambda d_r/Kp_y$ of the diffracted field. In the case where $(K, L) > (N, M)$, then $\rho_x > \Delta\xi$ and $\rho_y > \Delta\eta$. This means that the intrinsic resolution is not modified because it is imposed by the number of useful pixels (M,N) of the detector area and not by the number of data points of the reconstructed field. However, there is a decrease in the sampling pitch inducing an increase in the "definition" of the image plane. Definitively, this means that we will see more texture in the image: the resolution function will be finely sampled and the granular structure of the object will appear to the observer. Zero-padding of hologram consequently has to make the speckles of the image appear finely but without decreasing their size. This aspect is illustrated in Figure 1.12 in the case of a 2€ coin 25 mm in diameter illuminated at $\lambda = 632.8$ nm which was placed at distance d_0=660 mm from the sensor ($M \times N$ = 1,024 × 1,360 pixels, $p_x = p_y$ = 4.65 μm). The number of reconstructed points is chosen to be (512, 1,024, 2,048 and 4,096). When $K = L = 512$ there is a strong reduction of the width of the hologram since the number of data points used for the computation is smaller than the initial matrix. In this case, the intrinsic resolution decreases and the reconstructed image appears very badly. When $K = L = 1{,}024$, the number of data points is approximately equal to that given by the sensor (1,024 against 1,360 in horizontal direction). The image sampling also corresponds approximately to the intrinsic resolution. Thus, the image appears "pixelized". For $K = L =$ 2,048, zero-padding is effective and image sampling is twice as small as intrinsic resolution. So, the resolution function is sampled with a better definition and this facilitates the observation of the fine texture of the image, particularly its speckle. For $K = L = 4{,}096$, image sampling is now four times smaller than the intrinsic resolution. The definition of the image plane is again increased but the speckle does not change its size since it is imposed by the intrinsic resolution (i.e. ρ_x and ρ_y).

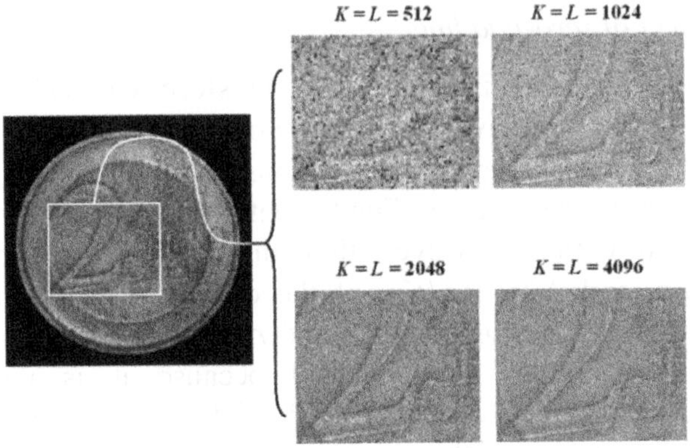

Figure 1.12. *Illustration of the effect of zero-padding*

1.3.2. *Reconstruction with convolution*

1.3.2.1. *Basic algorithm*

The numerical reconstruction with the convolution algorithm is based on the discrete version of equation [1.47] and considering the input plane as the hologram plane. In the same manner as that of the discrete Fresnel transform, the spatial sampling in the hologram plane has to be taken into account. Furthermore, we have to consider the sampling of the angular spectrum transfer function [1.34]. In the Fourier domain calculated by the FFT algorithm, the spatial frequencies are sampled so that $(u,v) = (n\delta u, m\delta v)$, where $(m;n) \in (-M/2,+M/2-1;-N/2,+N/2-1)$ and $(\delta u, \delta v) = (1/Np_x, 1/Mp_y)$.

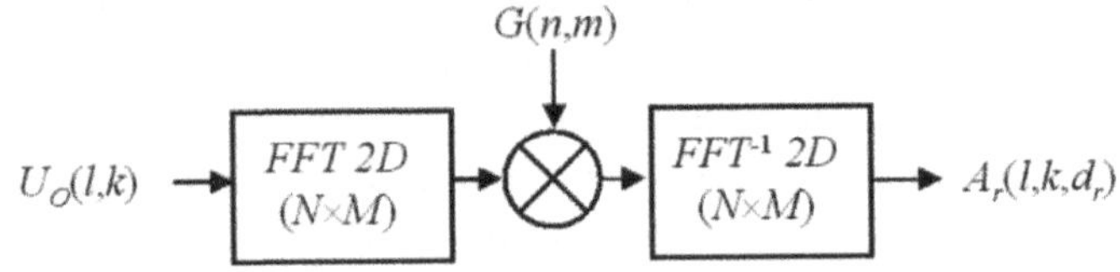

Figure 1.13. *Diagram of the reconstruction with convolution and the angular spectrum transfer function*

At any distance d_r from the recording plane, the reconstructed object field can be calculated according to the algorithm in

Figure 1.13, in which U_O is the object wave at the sensor plane estimated from the demodulation of the hologram (see section 1.1.3). This algorithm is known as the double-fast Fourier transform (D-FFT) algorithm since only two FFTs are required to compute the image plane.

Note that the sampling of the angular spectrum transfer function requires fulfilling the Shannon conditions. The condition is now:

$$d_r^{\max} \leq \min\left\{\frac{Lp_x^2}{\lambda}, \frac{Kp_y^2}{\lambda}\right\}, \qquad [1.51]$$

and is in the opposite way, compared to [1.48]. This means that the angular spectrum transfer function cannot be used to calculate distances larger than $d_r^{\max}$.

In cases of using directly the recorded hologram (i.e. H instead of U_O) as input to the algorithm in Figure 1.13, problems occur that are related to limitation in this approach. This point is discussed in the next section.

1.3.2.2. *Limits of classical approaches of convolution*

A spectral analysis is useful for highlighting the limits of the convolution approach when using directly H as input. From the convolution equation [1.40] and from equation [1.41], the impulse response of free space propagation is a bandwidth-limited function whose spatial bandwidth is related to the finite extent on which it is calculated. Thus, the spatial bandwidth of this convolution kernel is given by:

$$\Delta u_{\text{kernel}} \times \Delta v_{\text{kernel}} = \left(\frac{Np_x}{\lambda d_r}\right) \times \left(\frac{Mp_y}{\lambda d_r}\right) \qquad [1.52]$$

Also note that the transfer function of the convolution is a band-pass filter centered at (0,0) spatial frequency. From equation [1.12], the +1 order is localized at spatial frequencies (u_0,v_0). Thus, in order to reconstruct from H, by convolution, the image localized at spatial frequencies (u_0,v_0), we need to check the suitable frequency localization of the convolution kernel and the object-adapted spatial bandwidth. So as to fully reconstruct the object, the spatial frequency extents of the convolution kernel must be at least that of the object, giving:

$$\Delta u_{\text{kernel}} \times \Delta v_{\text{kernel}} \geq \Delta u_{\text{object}} \times \Delta v_{\text{object}} = \left(\frac{\Delta A_x}{\lambda d_r}\right) \times \left(\frac{\Delta A_y}{\lambda d_r}\right), \qquad [1.53]$$

where ΔA_x and ΔA_y are, respectively, the object width along the x- and y-directions. Equation [1.53] means that it is not possible to reconstruct directly an object by applying equation [1.47], whose size is larger than the width on which the impulse response is calculated (i.e. $Np_x \times Mp_y$). To increase the spatial frequency bandwidth of the convolution kernel, there are several strategies: (1) zero-padding of the impulse response or the angular spectrum transfer function, (2) design of a filter bank, which consists of a spectral scanning of the object spectrum so as to recover the full bandwidth, and (3) modifying the reconstruction distance so as to naturally increase the spatial bandwidth [PIC 13a]. Approaches 1 and 3 are described in the next two sections.

1.3.2.3. *Zero-padding of the impulse response*

A simple way to increase the spatial frequency bandwidth consists of using the zero-padding of the impulse response or the angular spectrum transfer function to get:

$$\Delta u_{\text{kernel}} = \frac{Lp_x}{\lambda d_r} = \Delta u_{\text{object}} = \frac{\Delta A_x}{\lambda d_r} \qquad [1.54]$$

leading to $L = \Delta A_x / p_x$. So, the number of data points is the ratio between the object size and the pixel pitch of the sensor. The problem linked to the reconstruction of extended objects by using a convolution approach is tantamount to a problem of adaptation of the spatial frequency bandwidth. In order to reconstruct the hologram, the transfer function may be the Fourier transform of the impulse response [1.41] or the angular spectrum transfer function [1.34]. For these options, we must take into account the spectral localization and the spatial frequency extents of the bandwidth. The zero-padding of the digital hologram with $(K,L)>(M,N)$ leads to extending the horizon for the calculation of the impulse response of the free space propagation, or to an oversampling of the angular transfer function. If the impulse response of free space propagation is used, the spectral localization can be obtained by using the modulation theorem. Centering at frequency (u_0,v_0) is performed by spatially modulating h, according to:

$$h_{\text{kernel}}(x,y,d_r)=h(x,y,d_r)\times\exp\left[+2i\pi(u_0x+v_0y)\right] \qquad [1.55]$$

If the angular spectrum transfer function is used, and as the transfer function must be a band-pass filter in the Fourier domain, we may limit the convolution kernel to the effective spectrum bandwidth and shift G (or G_F) in the spatial frequency space, and set a bandwidth limitation to satisfy sampling requirements. Thus, the associated transfer function becomes:

$$G_{\text{kernel}}(u,v,d_r)=\begin{cases}G(u-u_0,v-v_0,d_r) & \\ \quad \text{if } |u-u_0|\le Lp_x/2\lambda d_r \text{ and } |v-v_0|\le Kp_y/2\lambda d_r & \\ 0 \quad \text{elsewhere} & \end{cases} \qquad [1.56]$$

As an example, let us consider the numerical reconstruction of the digital hologram of an object sized $\Delta A_x = \Delta A_y = \Delta A = 60$ mm that is recorded with a pixel pitch at $p_x = p_y = 4.65$ µm. Given $L = \Delta A_x/p_x$, we need at least 12,000 × 12,000 data points to reconstruct the full object, which is greater than the calculation capacity of standard personnal computer (PC) computers. So, the approach based on the zero-padding remains limited to object having a "moderate" size, i.e. not too large compared to the sensor width.

1.3.2.4. *Adjustable magnification*

The second method to extend the spatial frequency bandwidth is based on modifying the reconstruction distance. For a given value of $\{K,L\}$, if d_r decreases, then the spatial bandwidth of equation [1.52] increases. The modification of the reconstruction distance can be obtained by using a spherical reconstruction wave, instead of a plane wave as a numerical wave impacting the hologram. This means that we need to consider H multiplied by a spherical wave $w(x, y, \lambda_c, R_c)$ (equation [1.25]).

The notion of "adjustable magnification" in digital holography requires a brief discussion. In classical holography, the concept of transverse magnification is perceptible since we can easily observe, before our very eyes, the change in the size and position of the diffracted image if we modify the laser wavelength or the curvature of the illuminating beam. However, with a numerical image, this notion is even less obvious. The notion of magnification is related to the field of view (sized $Lp_x \times Kp_y$) that is reconstructed by the convolution. The reconstruction of an extended object, whose physical size $\Delta A_x \times \Delta A_y$ is larger than the field of view, is expected. So that the object may fully appear in the horizon, a transversal magnification $\gamma = \min(Lp_x/\Delta A_x;Kp_y/\Delta A_y)$ must be applied; it is the ratio between the horizon and the object widths. In this way, we may conceive that the adjustable magnification method results in adjusting the reconstructed object size to that of the field of view, which is imposed by calculation capacities or computation speed. Note that several authors proposed algorithms qualified "with adjustable magnification". In 2004, Zhang and Yamaguchi [ZHA 04] proposed an algorithm based on a double Fresnel transform resulting in the adjustment of the side length of the field of view. In 2010, Restrepo and Garcia-Sucerquia [RES 10] discussed the adjustable magnification using the Fresnel–Bluestein transform. Both methods are based on the Fresnel transform (single or double) and the notion of magnification is linked to the ratio between the reconstructed pixel pitch and that of the sensor.

The transverse magnification is related to the ratio between reconstruction and recording distances [LI 09]:

$$\gamma = -\frac{d_r}{d_0}. \qquad [1.57]$$

Thus, the focus on the virtual image is not obtained for $d_r = -d_0$, but for a different distance that depends on the curvature of the recording reference wave and numerical reconstruction wave. This draws the basics of reconstruction with adjustable magnification: a spherical wave, either a reference or a reconstruction wave [LI 09, TAN 10, PIC 13a], will modify the reconstruction distance, which will modify the spatial frequency bandwidth of convolution, then leading to an adaptation to the object bandwidth. The consequence is that the reconstructed object gets a size which is now compatible with that of the reconstruction field of view. Note that $w(x,y,R_c)$ is also an oscillating function which must fulfill the sampling requirements to be correctly spatially sampled (see equation [1.48]).

The transfer function of convolution can be either the Fourier transform of the impulse response [1.41] or the angular spectrum transfer function (G or G_F) [1.34], [1.45]. For these options, we must take into account the spectral localization and the spatial frequency extents of the bandwidth. Since the transfer function is a band-pass filter in the Fourier domain, we may restrict it to the contour of the object and limit the convolution kernel to the effective object spectrum. Thus, if the object is included in a circular zone ($\Delta A_x = \Delta A_y = \Delta A$), the impulse response can be chosen to be:

$$h_{\text{kernel}}(x,y,d_r)=\begin{cases} h(x,y,d_r)\times\exp\left[+2i\pi(u_0x+v_0y)\right] \\ \qquad \text{if } x^2+y^2\leq\gamma^2\Delta A^2/4 \\ 0 \quad \text{elsewhere} \end{cases}. \qquad [1.58]$$

Similarly, the impulse response can also be defined for a rectangular object zone [PIC 09]. This restriction of the spatial bandwidth leads to a convolution kernel perfectly adjusted to the object bandwidth. If the angular spectrum transfer function is used, the associated transfer function is thus (circular object):

$$G_{\text{kernel}}(u,v,d_r)=\begin{cases} G(u-u_0,v-v_0,d_r) \\ \qquad \text{if } (u-u_0)^2+(v-v_0)^2\leq\gamma^2\Delta A^2/4 \\ 0 \quad \text{elsewhere} \end{cases}. \qquad [1.59]$$

Figure 1.14 shows the diagram for the algorithms based on convolution with adjustable magnification. This algorithm is known as the D-FFT algorithm since it uses two FFT computations

(Figure 1.14(a)) or as the three-fast Fourier transform (T-FFT) computation since it uses three FFT computations (Figure 1.14(b)).

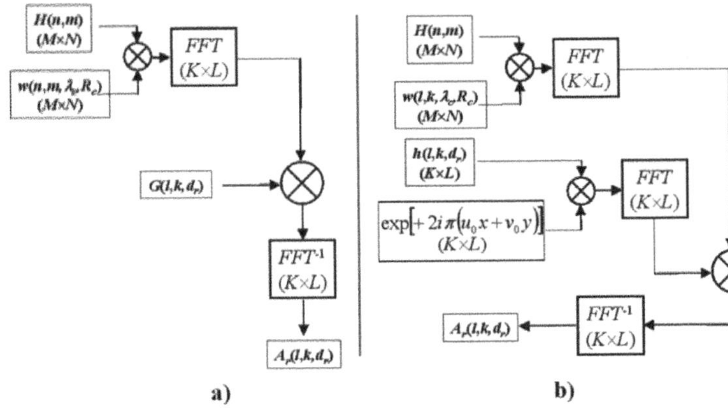

Figure 1.14. *Diagram of the reconstruction with convolution with adjustable magnification: a) with the angular spectrum transfer function and b) with the impulse response of free space propagation*

The use of the adjustable magnification induces a modification in the spatial resolution of the reconstruction process. Since the reconstruction distance changes due to the magnification, the spatial resolutions are:

$$\begin{cases} \rho'_x = \dfrac{\lambda d_r}{Np_x} = \gamma \dfrac{\lambda d_0}{Np_x} = \gamma\rho_x \\ \rho'_y = \dfrac{\lambda d_r}{Mp_y} = \gamma \dfrac{\lambda d_0}{Mp_y} = \gamma\rho_y \end{cases}, \qquad [1.60]$$

where ρ_x and ρ_y are the spatial resolutions of the holographic process given by the spatial extents of the recording area. From [1.60], the spatial resolution is proportional to the magnification. Since $|\gamma| < 1$ (object larger than sensor), this would mean that the spatial resolution is increased by the process. This is, of course, not physically possible because the numerical process cannot easily transcend the fundamental limits due to diffraction. As indicated previously, the reconstruction process based on adjustable magnification consists of modifying the object size in the direct space so that it can "enter" the window of the reconstructed field of view. Thus, it is interesting to note that the ratio between the object size and the spatial resolution remains unchanged, either in the direct space or in the reconstructed space. This means that the resolution is not increased in the magnified object. The only effect of the algorithm is on the image "definition", similarly to the effect of zero-padding in the Fresnel transform.

1.4. Holographic setups

This section aims to discuss the different experimental configurations to record a digital hologram. Basic experimental arrangements are provided and few explications on how it works are given.

1.4.1. *Fresnel holography*

The basic setup for digital Fresnel holography is given in Figure 1.15 [SCH 94]. The coherent light from a laser is separated into a reference wave and an object wave. The reference wave is expanded and illuminates directly the sensor, through the 50% beam splitter cube. The object wave, after being expanded, illuminates the object. The object diffracts light to the sensor area. Since the light is coherent, the mixing between the reference and the object waves produces interferences that constitute the digital hologram.

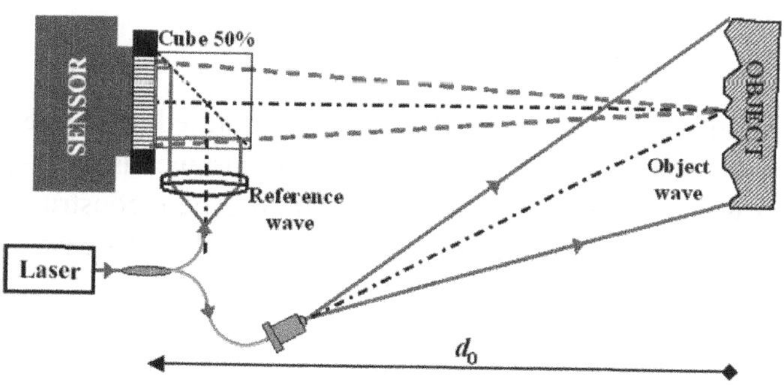

Figure 1.15. *Digital Fresnel holography. For a color version of this figure, see www.iste.co.uk/picart/digiholography.zip*

The numerical reconstruction can be performed with the discrete Fresnel transform in which the reconstruction distance is $d_r = -d_0$ if the reference wave is plane. The hologram can also be reconstructed using the convolution with adjustable magnification, and in this case $d_r \neq -d_0$.

1.4.2. *Fresnel holography with spatial spectrum reduction*

The basic setup is given in Figure 1.16. There are strong similarities with the digital Fresnel holography setup. The use of a diverging lens produces a virtual image of the object, which is smaller and localized closer to the sensor. So, this has for consequence to reduce the size of the setup and to give a more compact setup [SCH 96, MUN 10].

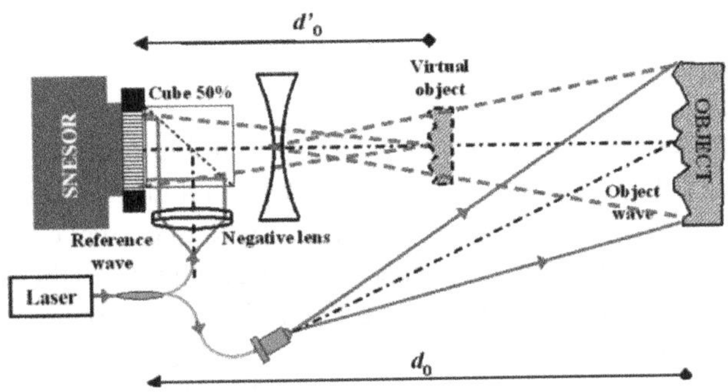

Figure 1.16. *Digital Fresnel holography with reduction of the spatial frequency spectrum. For a color version of this figure, see www.iste.co.uk/picart/digiholography.zip*

The numerical reconstruction can be performed with the discrete Fresnel transform in which the reconstruction distance is $d_r = -d'_0$ if the reference wave is plane. The hologram can also be reconstructed using the convolution with adjustable magnification, and in this case $d_r \neq -d'_0$.

1.4.3. *Fourier holography*

The basic setup is given in Figure 1.17. A convergent lens, with focal distance f', is inserted between the object plane and the sensor

[SEE 01]. The particularity is that the object and the sensor are localized at the focal points of this lens. It follows that the diffracted field at the sensor plane is simply the Fourier transform of the object [GOO 05].

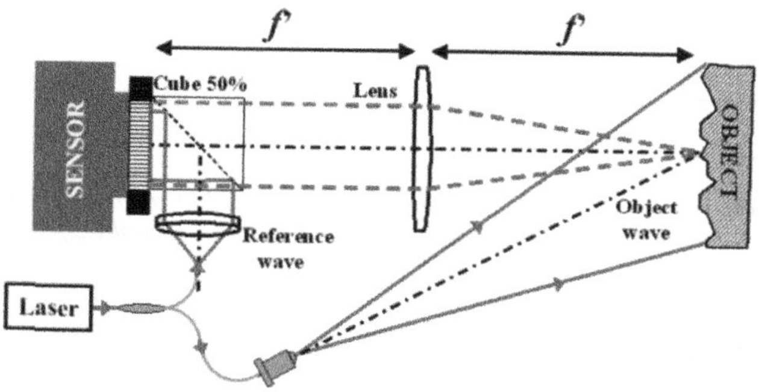

Figure 1.17. *Digital Fourier holography. For a color version of this figure, see www.iste.co.uk/picart/digiholography.zip*

The numerical reconstruction can be performed by computing directly the Fourier transform of the recorded hologram, without multiplying the hologram by the quadratic phase term. The main drawback of this setup is that the sensor and object have to be precisely adjusted at the focal points of the lens. In addition, the hologram of the optical mount can also be recorded in the hologram and this leads to parasitic images in the reconstructed plane.

1.4.4. *Lensless Fourier holography*

So as to simplify the Fourier setup, and to remove the lens, the lensless Fourier setup uses a spherical reference wave instead of a plane one [PED 02, ZHA 08]. The basic setup is given in Figure 1.18. The particularity is that the reference point source is localized in the object plane.

The quadratic terms in the Fresnel integral [1.43] compensate to give directly a Fourier transform. The object field can be reconstructed by calculating the inverse Fourier transform of the recorded hologram, equivalent to $d_r = \infty$ in algorithm of Figure 1.10.

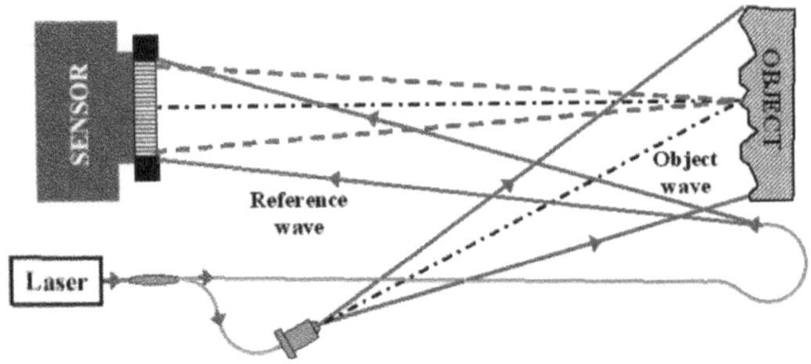

Figure 1.18. *Digital lensless Fourier holography. For a color version of this figure, see www.iste.co.uk/picart/digiholography.zip*

1.4.5. *Image-plane holography*

The basic setup is given in Figure 1.19. Consider an extended object having a size much larger than the recording area; image-plane holography projects the extended object near the sensor area and produces a reduction of the size of the image. A convergent lens, with focal distance f', is inserted between the object plane and the sensor, and it produces the reduced image of the object onto (or nearby) the sensor plane. In such a setup, the influence of the aperture diaphragm (AD) of the lens (AD in Figure 1.19) is significant. It must be pointed out that the recorded hologram is also the *digital Fresnel hologram of the aperture*. This means that the setup must be optimized according to the same rules as those for digital Fresnel holography. So, the NA of the imaging lens must be set to [KAR 12]:

$$\sin\alpha' \leq \frac{\lambda}{\left(2+3\sqrt{2}\right)p_x} \qquad [1.61]$$

If this relation is not fulfilled, the three diffraction orders of the hologram of the aperture will overlap. Thus, the useful +1 order of the object will be corrupted by the zero-order of the digital hologram of the aperture. Note that the NA depends only on the wavelength and the pixel pitch, whatever the object size may be, since the optimization of the setup is related to the aperture diameter. In the case of Fresnel holography, the useful NA of the beam is defined according to that of the sensor-to-object beam. It is equal to $\sin\alpha' \cong \Delta A/2d_0 = \lambda/(2+3\sqrt{2})p_x$ [PIC 08]. It follows that this relation is the same as [1.61]. From this standpoint, the optimization of the optical setup follows the same rules for both methods and does not depend on the object size. That has the consequence of the spatial resolutions of both methods being identical.

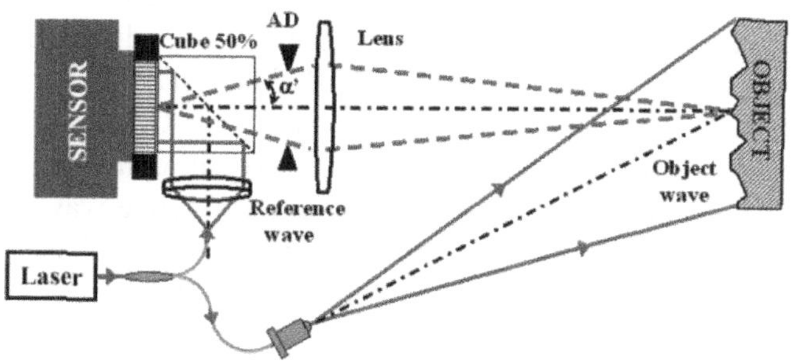

Figure 1.19. *Digital image-plane holography (AD: aperture diaphragm). For a color version of this figure, see www.iste.co.uk/picart/digiholography.zip*

From [1.48], the reconstruction distance d_r must fulfill the sampling condition of the quadratic phase in the discrete Fresnel transform. This means that, *a priori*, the image-plane hologram cannot be computed by the S-FFT method. So, the reconstruction is performed according to the D-FFT strategy with the angular spectrum transfer function, in which $d_r = 0$ when the object is rigorously projected in the recording plane. In the case of perfect image focusing, the transfer function of the convolution kernel tends to a uniform-bandwidth limited function [PED 95].

1.4.6. *Holographic microscopy*

Digital Holographic Microscopy (DHM) is the application of digital holography to microscopy. DHM is distinguished from other methods of microscopy by the fact that it does not require the focused recording of an image projected from the object onto the detector plane [GAB 49]. The object wave is simply projected by the objective of the microscope toward the detector plane. Since we are recording a digital hologram, we may reconstruct the object with a focus which can be digitally adjusted during the reconstruction process. As the refocusing is performed by digital means, it becomes possible to investigate in three-dimensional (3D) dynamical phenomena, which is actually not possible with usual light microscopy [DUB 06b, DUB 04b, DUB 06c, YOU 14]. There exist other, very similar, techniques of microscopy which differ in name, such as interferential microscopy, optical coherence tomography and diffraction phase microscopy [CUC 99b, DUB 99, DUB 04a, DUB 06b, MAN 05, BHA 12, GIR 13, RAJ 14, FER 06, SHA 10, GAB 12, PAR 12, MAN 08, PIC 13b]. These methods each have in common the use of the coherent combination of an object and a reference wave, allowing the ultimate obtainment of an amplitude image and a phase image of the object. DHM set-ups implemented with spatial partial coherence light sources has permitted to reduce and improve the accuracy of both phase and intensity images [DUB 99, DUB 06b, DUB 04b, DUB 06c, YOU 14]. In traditional microscopy, the image of the object is projected onto the detector, and since there is no reference wave, the essential phase information is lost. DHM facilitates getting what is known as "quantitative phase microscopy" [CHA 07, CHA 06a, CUC 00, CUC 99a, CUC 99b, FER 06, MAN 05, ZHA 98].

Principally, there exist two architectures: using transmission and using reflection. Figure 1.20 describes the basic scheme, without detailing the transmission or reflection configurations. The objective is represented by its lens assembly and AD. In the transmission configuration, the object is illuminated by a collimated beam, and diffracts the light toward the aperture cone of the microscope objective. In the reflection configuration, the object is illuminated by a collimated beam which first passes through the microscope objective, and then diffracts/reflects the light toward the aperture cone of the microscope objective. The light, therefore, crosses the microscope objective twice. Compared to image-plane holography, the configuration is slightly different since we aim at imaging microscopic or nanometric objects.

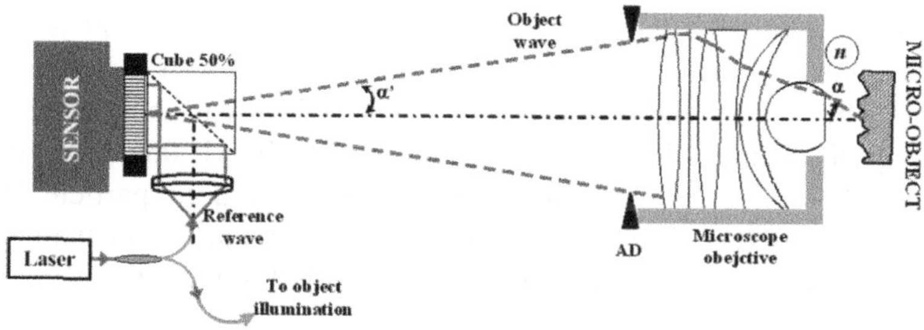

Figure 1.20. *Digital holographic microscopy. For a color version of this figure, see www.iste.co.uk/picart/digiholography.zip*

The image of the object near the sensor plane is produced by the microscope objective having a magnification $|\gamma_{opt}|>>1$ and a high NA $n\sin\alpha$. The holographic microscope has to be optimized similarly to image-plane holography. From the Abbe sinus relation and from [1.61], we have now the condition:

$$n\sin\alpha \leq \frac{|\gamma_{opt}|\lambda}{\left(2+3\sqrt{2}\right)p_x}. \qquad [1.62]$$

With pixels at a few microns and magnification larger than × 20, this relation is always fulfilled.

From [1.48], the reconstruction distance d_r must fulfill the sampling condition of the quadratic phase in the discrete Fresnel transform. This means that, as for image-plane holography, the processing cannot be carried out by the S-FFT method. The reconstruction is performed according to the D-FFT strategy with the angular spectrum transfer function.

1.4.7. *In-line Gabor holography*

In-line Gabor holography refers to the pioneering works of Gabor [GAB 48], who was studying an optical method to compensate for the aberrations in electron microscopy. The setup is quite simple and given in Figure 1.21. The laser beam is expanded and collimated to produce a parallel beam. It directly illuminates the sensor plane, and this beam constitutes the reference beam. In the path of the beam, an object diffracting light is inserted. This object can be opaque or slightly absorbing. Microscopy of phase objects [GAR 06a, GAR 06b] is also possible. In-line holography is well suited to particle field extraction [ONU 92, HIN 02] and to analyze their statistical size distribution [MAL 04, DEN 06], localization [COE 02, SIN 10] or 3D movement [HIN 02, VER 10].

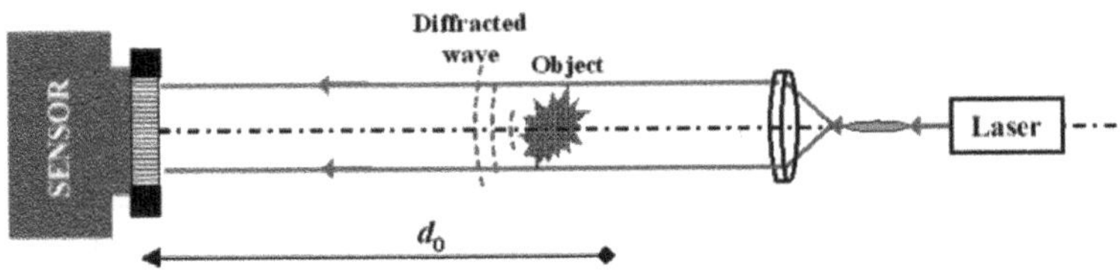

Figure 1.21. *Digital in-line Gabor holography. For a color version of this figure, see www.iste.co.uk/picart/digiholography.zip*

The modeling of the recorded hologram was provided by Onural [ONU 92, ONU 93]. The object distribution is denoted as $s(x,y)$ and the field emerging from the diffracting object plane when it is illuminated by a reference plane wave is denoted as $1-s(x,y)$. In-line holography generally considers that $s(x,y) \ll 1$. If the object distribution takes only real values, the in-line hologram can be approximately written as:

$$H(x,y) = 1 - s(x,y) * g_z(x,y), \qquad [1.63]$$

with g_z being given by:

$$g_z(x,y) = \frac{2}{\lambda z}\sin\left(\frac{\pi}{\lambda z}\left(x^2+y^2\right)\right). \qquad [1.64]$$

The reconstruction of the hologram can be performed using the discrete Fresnel transform, under the condition that the reconstruction distance is fulfilling equation [1.48], or with the fractional Fourier transform [COE 02]. The latter approach will not be detailed in this chapter (see Chapter 2).

Note that in-line holography, in its basic configuration, suffers from the overlapping of the twin image and zero-order, since there is no independent reference beam to produce a separation of the three orders. The use of spatial light modulator in the optical bath is a key to recover both the amplitude and phase in digital in-line holography [PIC 13b].

1.5. Digital holographic interferometry

Digital holographic interferometry exploits not only the amplitude of the object, but also its phase. This section presents the basic principle of digital holographic interferometry.

1.5.1. *Reconstruction of the phase of the object*

The result of the digital calculation of the diffracted field gives access to the complex amplitude sampled across a mesh corresponding to the number of reconstruction points of the algorithm. From this complex amplitude, we may access two quantities: the amplitude image (modulus) and the phase image (the argument of the field).The object phase, ψ_r, is obtained from:

$$\psi_r = \arg\{A_r\} = \tan^{-1}\left\{\frac{\Im m[A_r]}{\Re e[A_r]}\right\} \quad \mathrm{mod}(2\pi) \qquad [1.65]$$

The phase of the field is calculated using the arctangent function, and as a consequence, the result will be contained within the interval $[-\pi,+\pi]$, i.e. modulo 2π. This phase is random in most of the cases as it is related to the roughness of the object's surface. The reconstructed object is, therefore, marred with speckles [DAI 84]. The estimation of the optical phase of the reconstructed field is key to a large number of applications in digital holography. Note that the phase is relative to an unknown constant. Thus, the absolute phase cannot be obtained with a single wavelength. For the same reason, the notion of relief in digital holography is very different from that in analog holography. Several authors consider that digital holography, using a single hologram, as only reconstructing a "2.5D" volume [KOU 07a].

1.5.2. *Optical phase variations and the sensitivity vector*

Non-contact measurements using holographic methods are based on the variation of the optical phase of the reconstructed object when it is subjected to a stress. This stress may be of a biological, electronic, pneumatic, thermal, acoustic or mechanical nature. When subjected to a stress of any kind, the object is deformed, and thus the optical path along the *source-object-hologram* trajectory will vary. Let us imagine a point A at the light source and a point B attached to the object. When the object is slightly deformed by a stress, point B attached to the object undergoes a 3D change, the displacement vector $\mathbf{D}(D_x, D_y, D_z)$ which generates variations in the optical path from A to B and from B to C (Figure 1.22). These variations are much smaller than the absolute values of these path lengths, and have modules on the order of tens or hundreds of wavelengths of the used light. We denote by $\mathbf{K_e}$ the "illumination" vector of the object, by $\mathbf{K_o}$ the "observation" vector of the object and by n the refractive index around the object. The variation of optical path length is [KRE 96]:

$$\delta_{\text{opt}}(\text{ABC}) = n\mathbf{K}_\text{e}.\mathbf{D} - n\mathbf{K}_\text{o}.\mathbf{D} = n\mathbf{D}.(\mathbf{K}_\text{e} - \mathbf{K}_\text{o}) \quad [1.66]$$

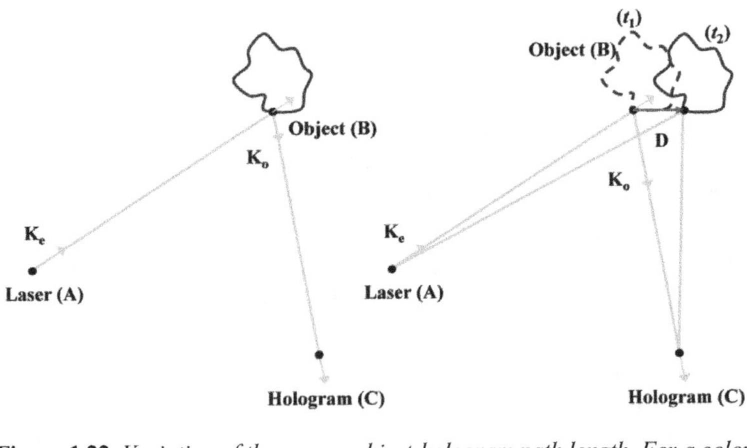

Figure 1.22. *Variation of the source-object-hologram path length. For a color version of this figure, see www.iste.co.uk/picart/digiholography.zip*

The observation vector is related to the direction of observation from the object toward the hologram. The illumination vector represents the direction of illumination of the studied object. We pose:

$$\mathbf{S} = \mathbf{K}_e - \mathbf{K}_o \tag{1.67}$$

This vector is called the *sensitivity vector*. The sensitivity vector corresponds to the difference between the illumination and observation vectors of the object. The sensitivity vector indicates the displacement direction in which the sensitivity of the apparatus is optimal. Knowledge of the coordinates of this vector is essential for the precise analysis of the amplitude of the displacements. The variation in optical phase induced by the variation in *source-object-hologram* optical path length is, therefore, given by the following relation [KRE 96]:

$$\Delta\phi = \frac{2\pi}{\lambda}\delta_{\text{opt}}(\text{ABC}) = \frac{2n\pi}{\lambda}\mathbf{D}.(\mathbf{K}_e - \mathbf{K}_o) \tag{1.68}$$

When the object is displaced along the displacement vector **D**, this leads to a variation of the phase, which is itself due to the variation in optical path.

1.5.3. *Phase difference method*

The measurement of the optical phase variations generated by the object requires the recording and reconstruction of at least two holograms (double exposure principle). The first corresponds to a reference hologram, and the second corresponds to a hologram of the object having been subjected to the change. Consequently, the phase variation may be evaluated by calculating the difference in optical phase between the two holograms. Let ψ_{r1} and ψ_{r2} be the optical phases of the first and second hologram, respectively. We then have:

$$\Delta\phi = \psi_{r2} - \psi_{r1} \quad \text{mod}(2\pi) \tag{1.69}$$

This phase variation will produce digital interference fringes, modulo 2π, which let us quantify the modification of the object between the two states. The variation in optical path seen in holographic interferometry, therefore, corresponds to the variation in position of the object projected onto the sensitivity vector. We note that for a large number of applications, the refractive index of the medium in which the studied object is placed often equals 1 (air), except for DHM in which cells may be living specimen.

As an illustration, Figure 1.23 shows two phases ψ_{r1} and ψ_{r2} of the first and second holograms, respectively, as well as the phase difference calculated modulo 2π. The two phases are random and uniformly distributed across $[-\pi,+\pi]$. The phase difference is also obtained in $[-\pi,+\pi]$. The case is that of a rough object surface. We observe digital interference fringes which represent phase jumps each time that $\Delta\varphi$ passes $-\pi$ or $+\pi$. We also observe that the result is noisy, which is translated by the appearance of a "salt-and-pepper" texture in the image. This noise is due to the decorrelation of the speckle pattern which exists more or less for each movement of the object [DAI 84, KAR 12].

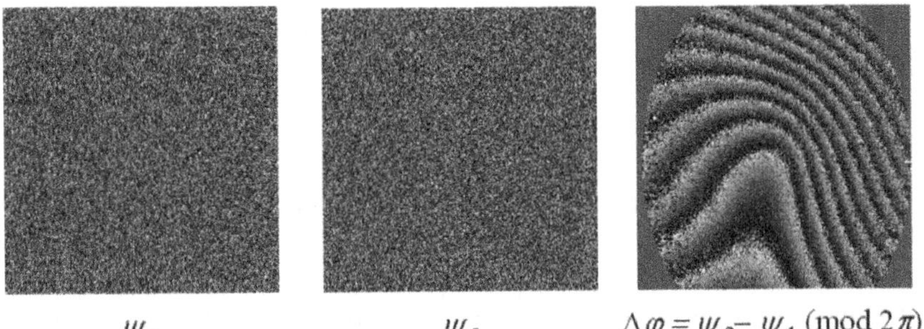

Figure 1.23. *Illustration of the phase-subtraction method*

Such results lead to two conclusions: it is necessary to spatially filter the result to reduce the level of noise (not discussed here) [AEB 99], and, it is necessary to reconstruct the continuity of the phase variation, which is destroyed by calculation using the arctangent function.

1.5.4. *Phase unwrapping*

This operation consists of reconstructing the physical continuity of the phase map. Of course, this continuity may only be reconstructed to within a phase constant, unless we know at which points of the mapping the phase is strictly zero. Figure 1.24 presents an illustration of this operation on one line of a result obtained after the subtraction of two phases.

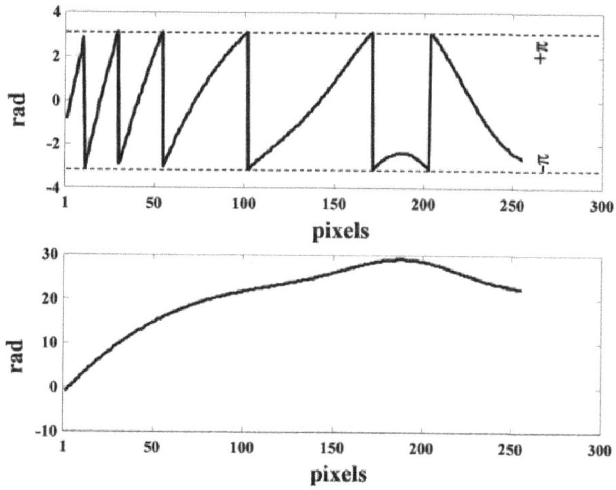

Figure 1.24. *Illustration of the "unwrapping" procedure*

The unwrapping algorithm starts at pixel N°1, and then searches for the phase jumps; at each one detected, -2π or $+2\pi$, it adds $+2\pi$, or -2π so that the phase of two neighboring points is continuous. Phase unwrapping techniques have become more sophisticated in recent years with the appearance of powerful algorithms whose implementation is not always straightforward. For a complete description, the readers should refer to [GHI 98].

1.6. Quantitative phase tomography

In general, a single wavefront determination obtained from a single hologram does not suffice to obtain a full 3D image of an object

[MAR 13]. An exception concerns sparse media, similar to those found in particle image velocimetry (PIV) where each particle can be considered as an isolated punctual object and its position determined in 3D by the methods of in-line holography. In this case, we can resort to sparse image representations to image the cloud of particles in 3D [LIE 04b, LIE 04c, DIX 11]. We must, however, admit that generally the true 3D shape of a given specimen cannot be derived from a single measurement obtained at one wavelength [SHE 10, KOU 07b]. The combination of data reconstructed from several holograms, obtained at either several wavelengths [YU 05] or at several incidence angles (multi-*k*), is needed to achieve true 3D imaging. Several articles report results obtained by changing the wavelength (variable *k*-vector amplitude) as shown, for example, by Marron [MAR 93]. However, in this study, the range of wavelength scan remains very small, and consequently the resolution is weak. Arons *et al.* using Fourier synthesis holography [ARO 96] have also discussed a similar approach. In digital holography, multiple wavelengths have been used to reconstruct 3D structures [KIM 99].

In the field of microscopy, this tomographic imaging technique, based on reconstruction from multiple holograms, has yielded very accurate images of cells, erythrocytes in particular [KUH 09, MON 06]. An alternative, but somehow similar approach, is to use a wide bandwidth source and form a hologram in the plane where the mutual coherence between the object and reference waves is non-zero: this concept introduces coherence gating in the space domain. It has proved to perform well [CUC 97, MAS 05].

The approach consisting of varying the angle of the illumination waves (variable *k*-vector direction) can be used in conjunction with the previous technique where the wavelength is changed (variable *k*-vector amplitude). This angular approach more exactly meets the concept found in the literature under the name of "diffraction tomography" [WOL 69] for reconstruction of the scattering potential associated with the structure of the diffracting object. A diffracted wave can be collected and reconstructed from the holograms at various incidences. The complex amplitude of the measured scattered field is linked to the object function via their Fourier transforms. In microscopy, a tomographic approach, not based on holography, was presented by Lauer [LAU 02]. A simple way to reconstruct the scattered wave can be based on digital holography: the phase and amplitude of the diffracted wave are directly reconstructed from the hologram and can be used to compute the scattering potential at every point of the specimen.

The scalar approximation has been shown to give good results. However, note that under the first Born approximation [WOL 69], the reconstruction is strictly valid for a low-phase change only, which is proportional to both the object refractive index and size, while under the Rytov approximation, "the size of the object is not a factor" (from [SLA 84]) and therefore the Rytov approximation is expected to give better results for larger objects.

Noting the wave vector of the diffracted wave k_d, k_i that for the illumination wave, and k_O that for the object field, these quantities are linked by the following equation:

$$k_d = k_i + k_O \quad [1.70]$$

When considering a transmission geometry and only one direction of illumination (only one k_i), only one half of the Ewald sphere could be recorded at best. Moreover, this half-sphere is restricted to a cap of sphere only, because of the limited NA of the microscope objective (NA) used in the holographic scheme (see Figure 1.20). As shown in Figure 1.25(a), a very limited subset of the diffracted wave vectors can be measured. From [1.70], for a given incident direction k_i, and an observation direction k_d, the corresponding wave vector of the object field is $k_O = k_d - k_i$. The set of wave vectors $k_d - k_i$ which are recorded is shown as a bold arc of circle in Figure 1.25(a). The recorded spatial frequency space can be filled using successive variable directions of illumination, leading to recording a more complete subset of the 3D frequency representation of the object, in order to perform a more accurate reconstruction of the object [DEB 08]. The use of tilted illuminating wave permits to record higher frequencies from the Fourier transform of the object field. Figure 1.25(b) illustrates the mapping of the k-space with the set of wave vectors of Figure 1.25(a).

Each circle represents the spatial bandwidth available from the holographic setup for each illumination vector. Figure 1.25(c) shows that after scanning the incident vector, a portion of the *k*-space is covered by the measured data. When a large number of incidences angles are scanned, the area covered by the measured wave vectors in the *x*-*y* plane becomes a disk (Figure 1.25(d)).

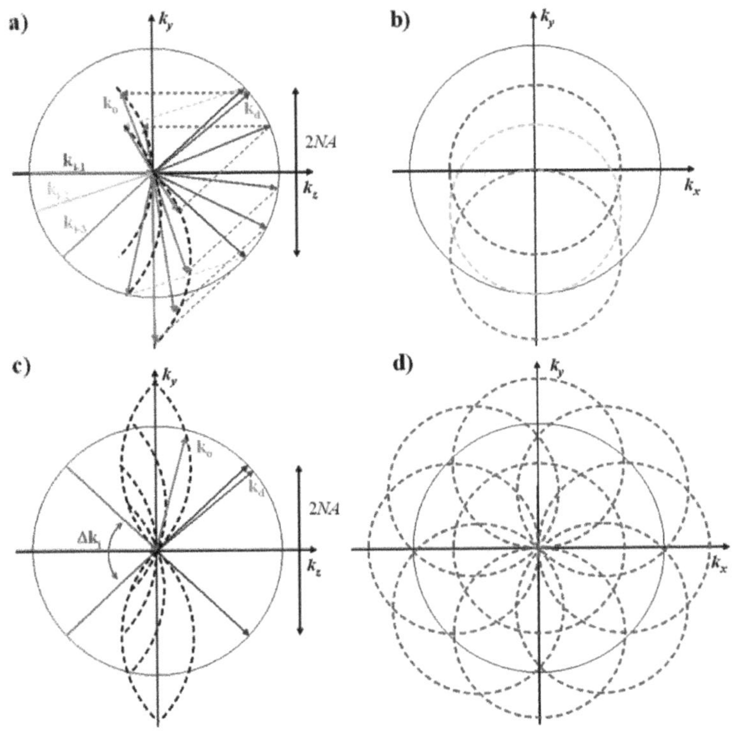

Figure 1.25. *Principle of diffractive tomography: construction of the measured spatial frequency support in the case of transmission (refer to text). For a color version of this figure, see www.iste.co.uk/picart/digiholography.zip*

From these data, the 3D distribution of the refractive index can be established providing, therefore, a 3D tomographic image of the cell constituents.

Tomography of cells based on DHM is new and original. In 2006, a first approach consisting of the rotation of the specimen has been developed [CHA 06b, CHA 06c]. It was demonstrated that the refractive index of the cell body could be measured in 3D with a spatial resolution better than 3 µm in all directions. Other approaches consisting of rotating the beam rather than the object have been proposed [DEB 09, CHO 07, SUN 09, KIM 11, COT 11]. Kou [KOU 07b] and Sheppard [SHE 10] have compared both approaches in their principles and shown consistent differences in the performance of each modality. In this case, the approach is commonly described as "synthetic aperture imaging" because the effective aperture results from the stitching of several loci (Ewald spheres) in the Fourier domain corresponding to various illuminating directions.

1.7. Conclusion

In this chapter, we have introduced the principles of digital holography, as well as the associated reconstruction methods. Some demodulation techniques of eliminating the zero-order and twin image, to increase the spatial resolution of reconstructed images, have been discussed. The discrete Fresnel transform is the simplest and the most direct calculation method, on the condition that the sampling theorem is respected in the plane of the hologram. This method is certainly the most widely used method, even though it does not guarantee the invariance of the reconstructed horizon with the wavelength [FER 04]. We have presented the methods based on convolution, especially the adjustable-magnification approach which leads to the adaptation of the reconstructed image size to the size of the field of view.

Note that other reconstruction methods based on wavelets are also discussed in the literature. For example, the readers could have a look at [ONU 93, LIE 03, LIE 04a] and [LIE 04b].

We have introduced the methods of holographic interferometry for which the phase of the reconstructed field is the principal parameter. The exploitation of the phase gives versatility to digital holography: quantitative microscopy, phase contrast, profilometry, the measurement of displacement fields, vibrations and fluid mechanics.

Quantitative phase tomography has been introduced and applications are detailed in Chapter 5.

This chapter serves as a basis for the next chapters that will go into advanced holographic methods more thoroughly.

1.8. Bibliography

[ABS 10] ABSIL E., TESSIER G., GROSS M., *et al.*, "Photothermal heterodyne holography of gold nanoparticles", *Optics Express*, vol. 18, pp. 780–786, 2010.

[AEB 99] AEBISCHER H.A., WALDNER S., "A simple and effective method for filtering speckle-interferometric phase fringe patterns", *Optics Communications*, vol. 162, pp. 205–210, 1999.

[ATL 07] ATLAN M., GROSS M., ABSIL E., "Accurate phase-shifting digital interferometry", *Optics Letters*, vol. 32, pp. 1456–1458, 2007.

[ATL 08] ATLAN M., GROSS M., DESBIOLLES P., *et al.*, "Heterodyne holographic microscopy of gold particles", *Optics Letters*, vol. 33, pp. 500–502, 2008.

[ARO 96] ARONS E., DILWORTH D., "Lensless imaging by spatial Fourier synthesis holography", *Applied Optics*, vol. 35, pp. 777–781, 1996.

[AWA 06a] AWATSUJI Y., FUJII A., KUBOTA T., *et al.*, "Parallel three-step phase-shifting digital holography", *Applied Optics*, vol. 45, pp. 2995–3002, 2006.

[AWA 06b] AWATSUJI Y., SASADA M., FUJII A., *et al.*, "Scheme to improve the reconstructed image in parallel quasi-phase-shifting digital holography", *Applied Optics*, vol. 45, pp. 968–974, 2006.

[BHA 12] BHADURI B., PHAM H., MIR M., *et al.*, "Diffraction phase microscopy with white light", *Optics Letters*, vol. 37, pp. 1094–1096, 2012.

[BOR 99] BORN M., WOLF E., *Principles of Optics*, 7th ed., Cambridge University Press, 1999.

[CHA 06a] CHARRIÈRE F., KÜHN J., COLOMB T., *et al.*, "Characterisation of micro lens by digital holographic microscopy", *Applied Optics*, vol. 45, pp. 829–835, 2006.

[CHA 06b] CHARRIÈRE F., MARIAN A., MONTFORT F., *et al.*, "Cell refractive index tomography by digital holographic microscopy", *Optics Letters*, vol. 31, pp. 178–180, 2006.

[CHA 06c] CHARRIÈRE F., PAVILLON N., COLOMB T., *et al.*, "Living specimen tomography by digital holographic microscopy: morphometry of testate amoeba", *Optics Express*, vol. 14, pp. 7005–7013, 2006.

[CHA 07] CHALUT K., BROWN W., WAX A., "Quantitative phase microscopy with asynchronous digital holography", *Optics Express*, vol. 15, pp. 3047–3052, 2007.

[CHO 07] CHOI W., FANG-YEN C., BADIZADEGAN K., *et al.*, "Tomographic phase microscopy", *Nature Methods*, vol. 4, pp. 717–719, 2007.

[COE 02] COETMELLEC S., LEBRUN D., OSKUL C., "Application of the two-dimensional fractional-order Fourier transformation to particle field digital holography", *Journal of the Optical Society of America A*, vol. 19, pp. 1537–1546, 2002.

[COL 70] COLLINS S.A., "Laser-system diffraction integral written in terms of matrix optics", *Journal of the Optical Society of America*, vol. 60, p. 1168, 1970.

[COL 06a] COLOMB T., MONTFORT F., KÜHN J., *et al.*, "Numerical parametric lens for shifting, magnification and complete aberration compensation in digital holographic microscopy", *Journal of the Optical Society of America A*, vol. 23, pp. 3177–3190, 2006.

[COL 06b] COLOMB T., CUCHE E., CHARRIÈRE F., *et al.*, "Automatic procedure for aberration compensation in digital holographic microscopy and applications to specimen shape compensation", *Applied Optics*, vol. 45, pp. 851–863, 2006.

[COQ 95] COQUOZ O., CONDE R., TALEBLOU F., *et al.*, "Performances of endoscopic holography with a multicore optical-fiber", *Applied Optics*, vol. 34, pp. 7186–7193, 1995.

[COT 11] COTTE Y., TOY M.F., DEPEURSINGE C., "Beyond the lateral resolution limit by phase imaging", *Journal of Biomedical Optics*, vol. 16, p. 106007, 2011.

[CRE 88] CREATH K., "Phase measurement interferometry techniques", in WOLF E. (ed.), *Progress in Optics*, North-Holland publishing Company, vol. 26, pp. 349–393, 1988.

[CUC 97] CUCHE E., POSCIO P., DEPEURSINGE C., "Optical tomography by means of a numerical low-coherence holographic technique", *Journal of Optics-Nouvelle Revue d'Optique*, vol. 28, pp. 260–264, 1997.

[CUC 99a] CUCHE E., MARQUET P., DEPEURSINGE C., "Simultaneous amplitude-contrast and quantitative phase-contrast microscopy by numerical reconstruction of Fresnel off-axis holograms", *Applied Optics*, vol. 38, pp. 6994–7001, 1999.

[CUC 99b] CUCHE E., BEVILACQUA F., DEPEURSINGE C., "Digital holography for quantitative phase contrast imaging", *Optics Letters*, vol. 24, pp. 291–293, 1999.

[CUC 00] CUCHE E., MARQUET P., DEPEURSINGE C., "Spatial filtering for zero-order and twin-image elimination in digital off-axis holography", *Applied Optics*, vol. 39, pp. 4070–7075, 2000.

[DAI 84] DAINTY J.C., *Laser Speckle and Related Phenomena*, Springer-Verlag, Berlin, 1984.

[DEB 08] DEBAILLEUL M., SIMON B., GEORGES V., *et al.*, "Holographic microscopy and diffractive microtomography of transparent samples", *Measurement Science and Technology*, vol. 19, pp. 074009–074016, 2008.

[DEB 09] DEBAILLEUL M., GEORGES V., SIMON B., *et al.*, "High-resolution three-dimensional tomographic diffractive microscopy of transparent inorganic and biological samples", *Optics Letters*, vol. 34, pp. 79–81, 2009.

[DEM 03] DEMOLI N., VUKICEVIC D., TORZYNSKI M., "Dynamic digital holographic interferometry with three wavelengths", *Optics Express*, vol. 11, pp. 767–774, 2003.

[DEN 06] DENIS L., FOURNIER C., FOURNEL T., *et al.*, "Direct extraction of the mean particle size from a digital hologram", *Applied Optics*, vol. 45, pp. 944–952, 2006.

[DES 08] DESSE J.M., PICART P., TANKAM P., "Digital three-color holographic interferometry for flow analysis", *Optics Express*, vol. 16, pp. 5471–5480, 2008.

[DES 11] DESSE J.M., PICART P., TANKAM P., "Sensor influence in digital 3λ holographic interferometry", *Measurement Science and Technology*, vol. 22, p. 064005, 2011.

[DES 12] DESSE J.M., PICART P., TANKAM P., "Digital color holography applied to fluid mechanics and structure mechanics", *Optics and Lasers in Engineering*, vol. 50, pp. 18–28, 2012.

[DIX 11] DIXON L., CHEONG F.C., GRIER D.G., "Holographic deconvolution microscopy for high-resolution particle tracking", *Optics Express*, vol. 19, pp. 16410–16417, 2011.

[DOR 99] DORRIO B.V., FERNANDEZ J.L., "Phase evaluation methods in whole-field optical measurement techniques", *Measurement Science and Technology*, vol. 10, pp. 33–55, 1999.

[DUB 99] DUBOIS F., JOANNES L., LEGROS J.C., "Improved three-dimensional imaging with a digital holography microscope with a source of partial spatial coherence", *Applied Optics*, vol. 38, pp. 7085–7094, 1999.

[DUB 04a] DUBOIS F., YOURASSOWSKY C., MONNOM O., "Microscopie en holographie digitale avec une source partiellement cohérente", in FAUPEL M., SMIGIELSKI P., GRZYMALA R. (eds.), *Imagerie et Photonique pour les sciences du vivant et la médecine*, Fontis Media, Fomartis, pp. 287–302, 2004.

[DUB 04b] DUBOIS F., REQUENA M.-L.N., MINETTI C., *et al.*, "Partial spatial coherence effects in digital holographic microscopy with a laser source", *Applied Optics*, vol. 43, pp. 1131–1139, 2004.

[DUB 06a] DUBOIS F., SCHOCKAERT C., CALLENS N., *et al.*, "Focus plane detection criteria in digital holography microscopy by amplitude analysis", *Optics Express*, vol. 14, pp. 5895–5908, 2006.

[DUB 06b] DUBOIS F., YOURASSOWSKY C., MONNOM O., *et al.*, "Digital holographic microscopy for the three-dimensional dynamic analysis of in vitro cancer cell migration", *Journal of Biomedical Optics*, vol. 11, pp. 054032.1–054032.5, 2006.

[DUB 06c] DUBOIS F., CALLENS N., YOURASSOWSKY C., *et al.*, "Digital holographic microscopy with reduced spatial coherence for three-dimensional particle flow analysis", *Applied Optics*, vol. 45, pp. 864–871, 2006.

[DUB 12] DUBOIS F., YOURASSOWSKY C., "Full off-axis red-green-blue digital holographic microscope with LED illumination", *Optics Letters*, vol. 37, pp. 2190–2192, 2012.

[FER 04] FERRARO P., DE NICOLA S., COPPOLA G., *et al.*, "Controlling image size as a function of distance and wavelength in Fresnel-transform reconstruction of digital holograms", *Optics Letters*, vol. 29, pp. 854–856, 2004.

[FER 05] FERRARO P., GRILLI S., ALFIERI D., *et al.*, "Extended focused image in microscopy by digital holography", *Optics Express*, vol. 13, pp. 6738–6749, 2005.

[FER 06] FERRARO P., ALFERI D., DE NICOLA S., *et al.*, "Quantitative phase-contrast microscopy by a lateral shear approach to digital holographic image reconstruction", *Optics Letters*, vol. 31, pp. 1405–1407, 2006.

[GAB 48] GABOR D., "A new microscopic principle", *Nature*, vol. 161, pp. 777–778, 1948.

[GAB 49] GABOR D., "Microscopy by reconstructed wavefronts", *Proceedings of the Royal Society*, vol. A1947, p. 454, 1949.

[GAB 12] GABAI H., SHAKED N.T., "Dual-channel low-coherence interferometry and its application to quantitative phase imaging of fingerprints", *Optics Express*, vol. 20, pp. 26906–26912, 2012.

[GAR 06a] GARCIA-SUCERQUIA J., XU W., JERICHO M.H., *et al.*, "Immersion digital in-line holographic microscopy", *Optics Letters*, vol. 31, pp. 1211–1213, 2006.

[GAR 06b] GARCIA-SUCERQUIA J., XU W., JERICHO S.K., *et al.*, "Digital in-line holographic microscopy", *Applied Optics*, vol. 45, pp. 836–850, 2006.

[GHI 98] GHIGLIA D.C., PRITT M.D., *Two-dimensional Phase Unwrapping: Theory, Algorithms and Software*, Wiley, New York, 1998.

[GIR 13] GIRSHOVITZ P., SHAKED N.T., "Compact and portable low-coherence interferometer with off-axis geometry for quantitative phase microscopy and nanoscopy", *Optics Express*, vol. 21, pp. 5701–5714, 2013.

[GOO 67] GOODMAN J.W., LAWRENCE R.W., "Digital image formation from electronically detected holograms", *Applied Physics Letters*, vol. 11, pp. 77–79, 1967.

[GOO 85] GOODMAN J.W., *Statistical Optics*, Wiley, New York, 1985.

[GOO 05] GOODMAN J.W., *Introduction to Fourier Optics,* 3rd ed., Roberts & Company Publishers, Greenwood Village, 2005.

[GOO 07] GOODMAN J.W., *Speckle Phenomena in Optics*, Ben Roberts and Co, Swansea, 2007.

[GRE 84] GREIVENKAMP J.E., "Generalized data reduction for heterodyne interferometry", *Optical Engineering*, vol. 23, pp. 350–352, 1984.

[GRO 07] GROSS, M., ATLAN M., "Digital holography with ultimate sensitivity", *Optics Letters*, vol. 32, pp. 909–911, 2007.

[GRO 08] GROSS M., ATLAN M., ABSIL E., "Noise and aliases in off-axis and phase-shifting holography", *Applied Optics*, vol. 47, pp. 1757–1766, 2008.

[HAR 02] HARIHARAN P., *Basics of Holography*, Cambridge University Press, New York, 2002.

[HIN 02] HINSCH K.D., "Holographic particle image velocimetry", *Measurement Science and Technology*, vol. 13, pp. R61–R72, 2002.

[HUA 71] HUANG T.S., "Digital holography", *Proceedings of the IEEE*, vol. 159, pp. 1335–1346, 1971.

[KAR 12] KARRAY M., SLANGEN P., PICART P., "Comparison between digital Fresnel holography and digital image-plane holography: the role of the imaging aperture", *Experimental Mechanics*, vol. 52, pp. 1275–1286, 2012.

[KHM 08] KHMALADZE A., MARTÍNEZ A.R., KIM M., *et al.*, "Simultaneous dual wavelength reflection digital holography applied to the study of the porous coal samples", *Applied Optics*, vol. 47, pp. 3203–3210, 2008.

[KIM 99] KIM M.K., "Wavelength-scanning digital interference holography for optical section imaging", *Optics Letters*, vol. 24, pp. 1693–1695, 1999.

[KIM 11] KIM M., CHOI Y., FANG-YEN C., *et al.*, "High-speed synthetic aperture microscopy for live cell imaging", *Optics Letters*, vol. 36, pp. 148–150, 2011.

[KOU 07a] KOU S.S., SHEPPARD C.J., "Imaging in digital holographic microscopy", *Optics Express*, vol. 15, pp. 13640–13648, 2007.

[KOU 07b] KOU S.S., SHEPPARD C.J.R., "Comparison of three dimensional transfer function analysis of alternative phase imaging methods", *Biomedical Optics (BiOS)*, pp. 64430Q–64430Q-6, 2007.

[KRE 86] KREIS T., "Digital holographic interference-phase measurement using the Fourier-transform method", *Journal of the Optical Society of America A*, vol. 3, pp. 847–855, 1986.

[KRE 96] KREIS T., *Holographic Interferometry – Principles and Methods*, Akademie Verlag Gmbh, Berlin, 1996.

[KRE 97] KREIS T., ADAMS M., JÜPTNER W., "Methods of digital holography: a comparison", *Proceedings of SPIE*, vol. 3098, pp. 224–233, 1997.

[KRE 04] KREIS T., *Handbook of Holographic Interferometry Optical and Digital Methods*, Wiley-VCH, Weinheim, 2004.

[KRO 72] KRONROD M.A., MERZLYAKOV N.S., YAROSLAVSKII L.P., "Reconstruction of a hologram with a computer", *Soviet Physics Technical Physics*, vol. 17, pp. 333–334, 1972.

[KUH 07] KUHN J., COLOMB T., MONTFORT F., *et al.*, "Real-time dual-wavelength digital holographic microscopy with a single hologram acquisition", *Optics Express*, vol. 15, pp. 7231–7242, 2007.

[KUH 08] KUHN J., CHARRIERE F., COLOMB T., *et al.*, "Axial sub-nanometer accuracy in digital holographic microscopy", *Measurement Science and Technology*, vol. 19, p. 044007, 2008.

[KUH 09] KUHN J., MONTFORT F., COLOMB T., *et al.*, "Submicrometer tomography of cells by multiple-wavelength digital holographic microscopy in reflection", *Optics Letters*, vol. 34, pp. 653–655, 2009.

[KUM 09] KUMAR U.P., BHADURI B., KOTHIYAL M.P., *et al.*, "Two-wavelength microinterferometry for 3-D surface profiling", *Optics and Lasers in Engineering*, vol. 47, pp. 223–229, 2009.

[LAN 08] LANGEHANENBERG P., KEMPER B., DIRKSEN D., *et al.*, "Autofocusing in digital holographic phase contrast microscopy on pure phase objects for live cell imaging", *Applied Optics*, vol. 47, pp. D176–D182, 2008.

[LAU 02] LAUER V., "New approach to optical diffraction tomography yielding a vector equation of diffraction tomography and a novel tomographic microscope", *Journal of Microscopy*, vol. 205, pp. 165–176, 2002.

[LAU 10] LAUTERBORN W., KURZ T., *Coherent Optics: Fundamentals and Applications*, Springer-Verlag, Berlin, 2010.

[LE 00] LE CLERC F., COLLOT L., GROSS M., "Numerical heterodyne holography with two-dimensional photodetector arrays", *Optics Letters*, vol. 25, pp. 716–718, 2000.

[LE 01] LE CLERC F., GROSS M., COLLOT L., "Synthetic-aperture experiment in the visible with on-axis digital heterodyne holography", *Optics Letters*, vol. 26, pp. 1550–1552, 2001.

[LEC 13] LECLERCQ M., KARRAY M., ISNARD V., *et al.*, "Evaluation of surface acoustic waves on the human skin using quasi-time-averaged digital Fresnel holograms", *Applied Optics*, vol. 52, pp. A136–A146, 2013.

[LEE 13] LEE K., KIM K., JUNG J., *et al.*, "Quantitative phase imaging techniques for the study of cell pathophysiology: from principles to applications", *Sensors*, vol. 13, pp. 4170–4191, 2013.

[LI 07] LI J.C., PENG Z., FU Y., "Diffraction transfer function and its calculation of classic diffraction formula", *Optics Communications*, vol. 280, pp. 243–248, 2007.

[LI 09] LI J.C., TANKAM P., PENG Z., *et al.*, "Digital holographic reconstruction of large objects using a convolution approach and adjustable magnification", *Optics Letters*, vol. 34, pp. 572–574, 2009.

[LIE 03] LIEBLING M., BLU T., UNSER M., "Fresnelets: new multiresolution wavelet bases for digital holography", *IEEE Transaction on Image Processing*, vol. 12, pp. 29–43, 2003.

[LIE 04a] LIEBLING M., On Fresnelets, interferences fringes, and digital holography, PhD Thesis, no. 2977, Ecole Polytechnique Fédérale de Lausanne, Switzerland, 2004.

[LIE 04b] LIEBLING M., BLU T., UNSER M., "Complex-wave retrieval from a single off-axis hologram", *Journal of the Optical Society of America A*, vol. 21, pp. 367–377, 2004.

[LIE 04c] LIEBLING M., UNSER M., "Autofocus for digital Fresnel holograms by use of a Fresnelet-sparsity criterion", *Journal of the Optical Society of America A*, vol. 21, pp. 2424–2430, 2004.

[MAL 04] MALEK M., ALLANO D., COËTMELLEC S., *et al.*, "Digital in-line holography: influence of the shadow density on particle field extraction", *Optics Express*, vol. 12, pp. 2270–2279, 2004.

[MAN 05] MANN C., YU L., CHUN-MIN L., *et al.*, "High-resolution quantitative phase-contrast microscopy by digital holography", *Optics Express,* vol. 13, pp. 8693–8698, 2005.

[MAN 08] MANN C.J., BINGHAM P.R., PAQUIT V.C., *et al.*, "Quantitative phase imaging by three wavelength digital holography", *Optics Express*, vol. 16, pp. 9753–9764, 2008.

[MAR 93] MARRON J.C., SCHROEDER K.S., "Holographic laser-radar", *Optics Letters*, vol. 18, pp. 385–387, 1993.

[MAR 05] MARQUET P., RAPPAZ B., MAGISTRETTI P., *et al.*, "Digital holographic microscopy: a non-invasive contrast imaging technique allowing quantitative visualization of living cells with sub-wavelength axial accuracy", *Optics Letters*, vol. 30, pp. 468–470, 2005.

[MAR 13] MARQUET P., DEPEURSINGE C., MAGISTRETTI P.J., "Exploring neural cell dynamics with digital holographic microscopy", *Annual Review of Biomedical Engineering*, vol. 15, pp. 407–431, 2013.

[MAS 99] MAS D., GARCIA J., FERREIRA C., *et al.*, "Fast algorithms for free-space diffraction patterns calculation", *Optics Communications*, vol. 164, pp. 233–245, 1999.

[MAS 03] MAS D., PEREZ J., HERNANDEZ C., *et al.*, "Fast numerical calculation of Fresnel patterns in convergent systems", *Optics Communications,* vol. 227, pp. 245–258, 2003.

[MAS 05] MASSATSCH P., CHARRIERE F., CUCHE E., *et al.*, "Time-domain optical coherence tomography with digital holographic microscopy", *Applied Optics*, vol. 44, pp. 1806–1812, 2005.

[MON 06] MONTFORT F., COLOMB T., CHARRIERE F., *et al.*, "Submicrometer optical tomography by multiple-wavelength digital holographic microscopy", *Applied Optics*, vol. 5, pp. 8209–8217, 2006.

[MUN 10] MUNDT J., KREIS T.M., "Digital holographic recording and reconstruction of large scale objects for metrology and display", *Optical Engineering*, vol. 49, pp. 125801-1–125801-6, 2010.

[ONU 92] ONURAL L., OZGEN M.T., "Extraction of three-dimensional object-location information directly from in-line holograms using Wigner analysis", *Journal of the Optical Society of America A*, vol. 9, pp. 252–260, 1992.

[ONU 93] ONURAL L., "Diffraction from a wavelet point of view", *Optics Letters*, vol. 18, pp. 846–848, 1993.

[PAR 12] PARTHASARATHY A.B., CHU K.K, FORD T.N., *et al.*, "Quantitative phase imaging using a partitioned detection aperture", *Optics Letters*, vol. 37, pp. 4062–4064, 2012.

[PED 95] PEDRINI G., TIZIANI H.J., "Digital double pulse holographic interferometry using Fresnel and image plane holograms", *Measurement*, vol. 18, pp. 251–260, 1995.

[PED 02] PEDRINI G., TIZIANI H.J., "Short-coherence digital microscopy by use of a lens less holographic imaging system", *Applied Optics*, vol. 41, pp. 4489–4496, 2002.

[PIC 03] PICART P., LEVAL J., MOUNIER D., *et al.*, "Time-averaged digital holography", *Optics Letters*, vol. 28, pp. 1900–1902, 2003.

[PIC 05] PICART P., LEVAL J., MOUNIER D., *et al.*, "Some opportunities for vibration analysis with time averaging in digital Fresnel holography", *Applied Optics*, vol. 44, pp. 337–343, 2005.

[PIC 08] PICART P., LEVAL J., "General theoretical formulation of image formation in digital Fresnel holography", *Journal of the Optical Society of America A*, vol. 25, pp. 1744–1761, 2008.

[PIC 09] PICART P., TANKAM P., MOUNIER D., *et al.*, "Spatial bandwidth extended reconstruction for digital color Fresnel holograms", *Optics Express*, vol. 17, pp. 9145–9156, 2009.

[PIC 12] PICART P., LI J.C., *Digital Holography*, ISTE, London and John Wiley & Sons, New York, 2012.

[PIC 13a] PICART P., TANKAM P., "Analysis and adaptation of convolution algorithms to reconstruct extended objects in digital holography", *Applied Optics*, vol. 52, pp. A240–A253, 2013.

[PIC 13b] PICART P., MALEK M., "Complex field recovering from in-line digital holography", *Optics Letters*, vol. 38, pp. 3230–3232, 2013.

[RAJ 14] RAJSHEKHAR G., BHADURI B., EDWARDS C., *et al.*, "Nanoscale topography and spatial light modulator characterization using wide-field quantitative phase imaging", *Optics Express*, vol. 22, pp. 3432–3438, 2014.

[RES 10] RESTREPO J.F., GARCIA-SUCERQUIA J., "Magnified reconstruction of digitally recorded holograms by Fresnel-Bluestein transform", *Applied Optics*, vol. 49, pp. 6430–6435, 2010.

[SAM 11] SAMSON B., VERPILLAT F., GROSS M., *et al.*, "Video-rate laser Doppler vibrometry by heterodyne holography", *Optics Letters*, vol. 36, pp. 1449–1451, 2011.

[SCH 94] SCHNARS U., JÜPTNER W., "Direct recording of holograms by a CCD target and numerical reconstruction", *Applied Optics*, vol. 33, pp. 179–181, 1994.

[SCH 96] SCHNARS U., KREIS T.M., JÜPTNER W., "Digital recording and numerical reconstruction of holograms: reduction of the spatial frequency spectrum", *Optical Engineering*, vol. 35, pp. 977–982, 1996.

[SCH 99] SCHEDIN S., PEDRINI G., TIZIANI H.J., *et al.*, "Simultaneous three-dimensional dynamic deformation measurements with pulsed digital holography", *Applied Optics*, vol. 38, pp. 7056–7062, 1999.

[SEE 01] SEEBACHER S., OSTEN W., BAUMBACH T., *et al.*, "The determination of material parameters of microcomponents using digital holography", *Optics and Lasers in Engineering*, vol. 36, pp. 103–126, 2001.

[SHA 10] SHAFFER E., MORATAL C., MAGISTRETTI P., *et al.*, "Label-free second-harmonic phase imaging of biological specimen by digital holographic microscopy", *Optics Letters*, vol. 35, pp. 4102–4104, 2010.

[SHE 10] SHEPPARD C.J.R., KOU S.S., "3D imaging with holographic tomography", *AIP Conference Proceedings*, vol. 1236, p. 65, 2010.

[SIE 86] SIEGMAN A.E., *Laser*, University Science Books, Sausalito Cal., 1986.

[SIN 10] SINGH D.K., PANIGRAHI P.K., "Improved digital holographic reconstruction algorithm for depth error reduction and elimination of out-of-focus particles", *Optics Express*, vol. 18, pp. 2426–2448, 2010.

[SLA 84] SLANEY M., KAK A.C., LARSEN L.E., "Limitations of imaging with first-order diffraction tomography", *IEEE Transactions on Microwave Theory and Techniques*, vol. 32, p. 860, 1984.

[SUN 09] SUNG Y.J., CHOI W., FANG-YEN C., *et al.*, "Optical diffraction tomography for high resolution live cell imaging", *Optics Express*, vol. 17, pp. 266–277, 2009.

[TAK 82] TAKEDA M., INA H., KOBAYASHI S., "Fourier-transform method of fringe-pattern analysis for computed-based topography and interferometry", *Journal of the Optical Society of America*, vol. 72, pp. 156–160, 1982.

[TAN 10a] TANKAM P., SONG Q., KARRAY M., *et al.*, "Real-time three-sensitivity measurements based on three-color digital Fresnel holographic interferometry", *Optics Letters*, vol. 35, pp. 2055–2057, 2010.

[TAN 10b] TANKAM P., PICART P., MOUNIER D., *et al.*, "Method of digital holographic recording and reconstruction using a stacked color image sensor", *Applied Optics*, vol. 49, pp. 320–328, 2010.

[TAN 11] TANKAM P., PICART P., "Use of digital color holography for crack investigation in electronic components", *Optics and Lasers in Engineering*, vol. 49, pp. 1335–1342, 2011.

[VER 10] VERRIER N., REMACHA C., BRUNEL M., *et al.*, "Micropipe flow visualization using digital in-line holographic microscopy", *Optics Express*, vol. 18, pp. 7807–7819, 2010.

[WOL 69] WOLF E, "Three-dimensional structure determination of semi-transparent objects from holographic data", *Optics Communications*, vol. 1, pp. 153–156, 1969.

[WYA 75] WYANT J.C., "Use of an ac heterodyne lateral shear interferometer with real-time wavefront correction systems", *Applied Optics*, vol. 14, pp. 2622–2626, 1975.

[YAM 97] YAMAGUCHI I., ZHANG T., "Phase-shifting digital holography", *Optics Letters*, vol. 22, pp. 1268–1270, 1997.

[YAM 01a] YAMAGUCHI I., KATO J., OHTA S., *et al.*, "Image formation in phase shifting digital holography and application to microscopy", *Applied Optics*, vol. 40, pp. 6177–6186, 2001.

[YAM 01b] YAMAGUCHI I., KATO J., OHTA S., "Surface shape measurement by phase shifting digital holography", *Optical Review*, vol. 8, pp. 85–89, 2001.

[YAM 02] YAMAGUCHI I., MATSUMURA T., KATO J., "Phase shifting color digital holography", *Optics Letters*, vol. 27, pp. 1108–1110, 2002.

[YAR 85] YARIV A., *Optical Electronics*, Holt, Rinehart and Winston, New York, 1985.

[YOU 14] YOURASSOWSKY C., DUBOIS F., "High throughput holographic imaging-in-flow for the analysis of a wide plankton size range", *Optics Express*, vol. 22, pp. 6661–6673, 2014.

[YU 05] YU L., KIM M.K., "Wavelength-scanning digital interference holography for tomographic three-dimensional imaging by use of the angular spectrum method", *Optics Letters*, vol. 30, pp. 2092–2094, 2005.

[ZHA 98] ZHANG T., YAMAGUCHI I., "Three-dimensional microscopy with phase shifting digital holography", *Optics Letters*, vol. 23, pp. 1221–1223, 1998.

[ZHA 04] ZHANG F., YAMAGUCHI I., "Algorithm for reconstruction of digital holograms with adjustable magnification", *Optics Letters*, vol. 29, pp. 1668–1670, 2004.

[ZHA 08] ZHAO J., JIANG H., DI J., "Recording and reconstruction of a color holographic image by using digital lensless Fourier transform holography", *Optics Express*, vol. 16, pp. 2514–2519, 2008.

2

Digital In-line Holography Applied to Fluid Flows

Due to the simplicity of its optical configuration, in-line holography offers an attractive way to study fluid flows [COË 01, PAN 01, SHE 03, FOU 04, GAR 06]. The use of a single beam is particularly interesting in industrial or laboratory flow studies where optical access is difficult. Then, this configuration opens the way for onboard measurement systems [OWE 00, FUG 04]. Generally, it can be said that for a given situation where classical transmission imaging is feasible, holography can also be applied. This chapter is organized as follows: two typical applications are described; the first application is concerned with near-wall velocity measurements and the second application is concerned with bubbles sizing in a cavitation tunnel. A secondary advantage of the in-line configuration is that the coherence of the light source is not as critical as it is for off-axis configuration. Then, theoretical models of holograms recorded with partial coherent sources (temporal or spatial) are described and the reconstruction of these holograms by fractional Fourier transform is presented. Finally, it must be said that the tracers generally encountered in flows (for example, bubbles or droplets) are not opaque. This chapter extends the application of digital in-line holography (DIH) to the determination of size and three-dimensional (3D) location of phase objects.

Chapter written by Sébastien COËTMELLEC, Denis LEBRUN and Marc BRUNEL.

2.1. Examples of measurements in flows

The following examples describe direct applications of DIH to particle characteristics measurements in flows. The first example presents the measurement of 3D velocity vector field in a turbulent boundary layer. In the second example, a similar setup is used to measure the size of micrometric bubbles in a cavitation tunnel. For both examples, a holographic magnification is required not only for enlarging particle images but also to increase the spatial resolution of reconstructed images. Through these examples, we will see hereafter how to position the different elements (laser source and camera) from the sample volume for recording hologram under optimal conditions.

2.1.1. *Increasing NA with a divergent wave*

The use of micrometric tracers needs to use an optical magnification. When the two-dimensional (2D) sensor (charge-coupled device (CCD) or complementary metal-oxide semiconductor (CMOS) camera) cannot be placed near the sample to be studied, an optical system is used in order to report particle images near the camera by using a microscope objective lens [SHE 08, KAT 06, STA 13, MAL 08]. However, the introduction of optical elements may lead to inconveniences, such as astigmatism. A magnification factor can also be easily obtained by illuminating the sample volume with a diverging beam. It must be said that a diverging illumination does not directly produce an improvement of the spatial resolution. Nevertheless, the interference fringes are spread over a larger area so that the image sensor can be brought closer to the objects to be recorded without violating the Shannon's criterion [LEB 11, KEL 11]. Consequently, the numerical aperture (NA) of the recording system can be better adapted to micrometric objects and a resulting significant change of the scaling is observed in the reconstructed images [XU 03, GAR 06]. The optical configuration used in the two following examples is given in Figure 2.1. A single laser beam coming from a point source S illuminates the objects and a 2D detector records the hologram.

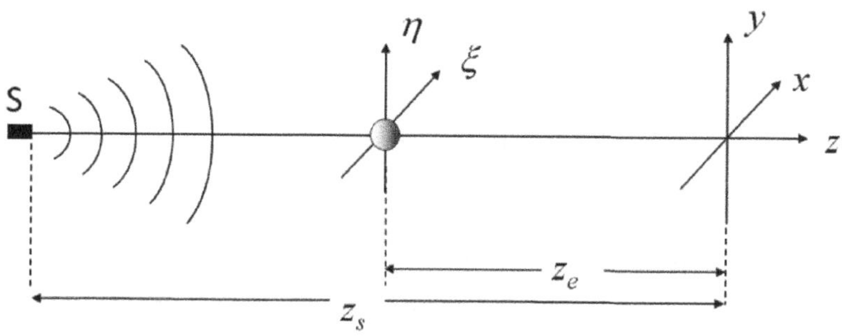

Figure 2.1. *Description of coordinates system*

The point source and the object are located, respectively, at distances z_s and z_e from the camera. According to [LEB 11], a magnification factor K, defined by:

$$K = \frac{z_s}{z_s - z_e} \quad [2.1]$$

can be efficiently introduced to describe the intensity distribution viewed by the sensor. Then, provided that the far-field conditions are fulfilled (i.e $\frac{\pi d^2}{4\lambda z_e} \ll 1$), the hologram pattern of a particle of diameter d centered on the optical axis can be written as follows [VIK 88]:

$$I_{z_e}(r) = 1 - 2\lambda z_{eq} \sin(\frac{\pi r^2}{\lambda z_{eq}}) F_{\frac{d_{eq}}{\lambda z_{eq}}}(r) + (\lambda z_{eq})^2 F^2_{\frac{d_{eq}}{\lambda z_{eq}}}(r) \quad [2.2]$$

where $F_\alpha(r) = \frac{\pi}{2}\alpha^2 \frac{J_1(\pi\alpha r)}{(\pi\alpha r)}$, $z_{eq} = K z_e$ and $d_{eq} = Kd$. By means of this equation, we can easily describe the intensity distribution of a particle hologram irrespective of the divergence of the recording beam. This change of scale also gives a direct method for hologram reconstruction because the fringe patterns are processed exactly as if they were produced in a plane wave configuration. Afterwards, a simple ratio of the measured coordinates and size by the magnification factor K has to be applied to recover the characteristics on the object field.

2.1.2. *Choice of the magnification*

From equation [2.1], we see that the magnification K is dependent on the distance of the object field to the CCD sensor z_e. This distance has to be minimized for increasing the NA necessary to obtain particle-images that can be detected and localized. However, z_e must be large enough such that the Shannon/Nyquist criterion be verified [LEB 11]. This implies that, for a camera with $N \times N$ pixels of size p, the inequality $z_e > \frac{Np^2}{K\lambda}$ must be checked. Moreover, other constraints such as the distance source-CCD sensor z_s have to be taken into account. From equation [2.1], expressing z_e as a function of z_s and K in this inequality leads to:

$$K > \frac{Np^2}{\lambda z_s} + 1 \qquad [2.3]$$

This inequality gives, for a given distance source-camera, the minimum magnification that could be used for safely recording holograms on a discrete sensor.

2.1.3. *3D velocity measurements in a turbulent boundary layer*

Velocity measurements in the vicinity of an obstacle remain very complicated even when optical diagnostics based on displacement of micrometric tracers are considered. Here, we show how it is possible to measure the 3D velocity vector fields in a turbulent boundary layer and, in particular, on the near-wall region of a wind tunnel [ALL 12]. In order to measure airflow velocity accurately, the size of the droplets (here used as tracers) should lie within the range of $1 - 5\ \mu m$ to ensure a minimum velocity lag between the tracers and the flow. Consequently, the NA of the recording system must be adapted to the present situation. The system developed for recording holograms in the wind tunnel is shown in Figure 2.2. The sample volume (air seeded with fog droplets near the window-reticle) is twice illuminated by a diverging beam coming from a single-mode optical fiber with a $5\ \mu m$ core. The mechanical stability of the system is ensured by inserting the fiber in a metallic needle. The needle-fiber holder, fixed on the wind tunnel downstream of the object field, is far enough from the investigation volume to avoid any disturbance of the flow. All the parameters of the experimental setup are summarized in the following array:

Parameters of the experimental setup	Value
Power of the laser source	25 mW
Wavelength	640 nm
CCD	Hamamatsu, C9300, 2048 $\times$ 2048, 12 bits
Pixel size	$7.4\ \mu m \times 7.4\ \mu$ m
Pulse duration	1.05–1.72 μ s
Depth of sample volume	1.25 mm
Magnification range	$3.19 \leq$ K ≤ 3.33
Near-wall object field	2.8 mm $\times$ 2.8 mm
Equivalent spatial sampling period	2.3μ m $\times$2.3 μ m

Note that we have observed that by choosing experimentally $z_s \simeq 80\ mm$, the sensor plane is well exposed by the laser beam. Using the characteristics of the experimental setup, we see from equation [2.3] that the magnification K should be at least equal to 3.1. Here, we have found that a magnification $K \simeq 3.2$ is a quite good compromise to reconstruct exploitable particle images.

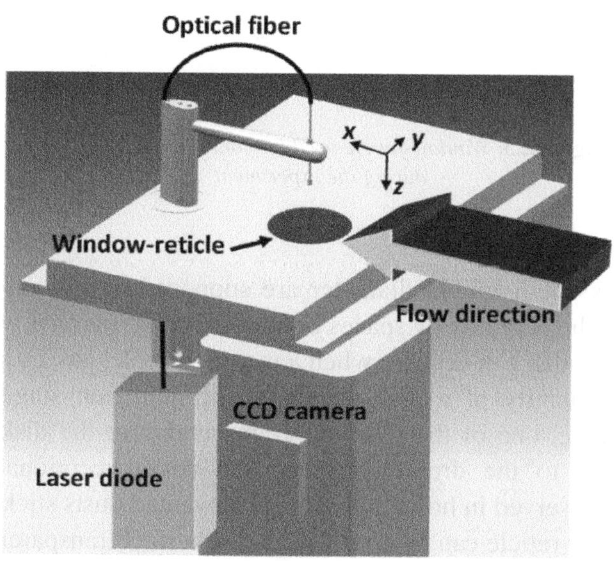

Figure 2.2. *Holographic probe for near-field velocity measurements*

2.1.3.1. *Recording holograms through a window-reticle*

The aim of the study consists of accurate measurements of particle velocities knowing their distance from the wall. As a result, it is crucial to localize most exactly as possible the $z - coordinate$ of the wall during the recording of a hologram sequence. For this purpose, all the holograms have been recorded through a transparent reticle specially designed for this application where five opaque disks are etched. These calibrated objects, spatially distributed as shown in Figure 2.3, are separated by certain distances, guaranteed by the manufacturer with an error smaller than $0.25\ \mu m$.

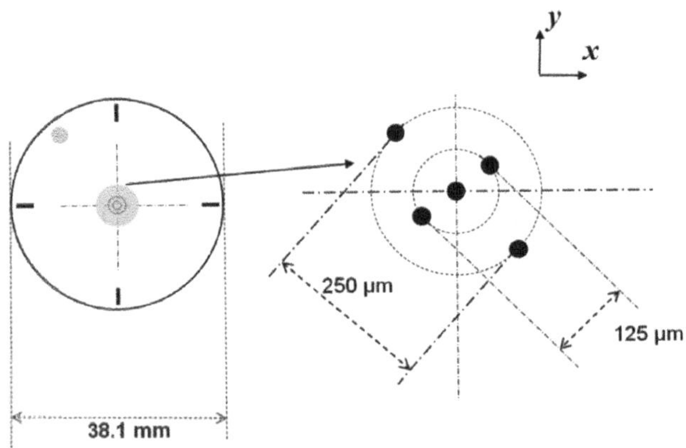

Figure 2.3. *Window-reticle used for continuous calibration during the experiment*

The disks of $5\ \mu m$ in diameter are supposed to optically disturb exclusively the very nearby spaces of measurement: in other words, the only problem that can occur is when a moving droplet passes just in the very close vicinity of a disk during the illumination stage. In this particular case, one of the two holograms is due to the disk and the other is due to the droplet. Except this very exceptional case – commonly observed in holography when unwanted dusts stick onto the window – our reticle can be considered as a perfect transparent optical window. Besides, as it can be seen in [ALL 12], the diffraction pattern produced by the window-reticle is completely removed by the normalization process.

Figure 2.4. *Reconstructed hologram of the window-reticle (five disks of diameter 5 μm)*

2.1.3.2. *Error estimations*

The intrinsic spatial resolution of digital holography has already been extensively studied in the past. It is roughly dependent on the NA of the hologram [KRE 05, PIC 08, ZHA 06]. Here, the estimation of 3D coordinates of image pairs may be hampered by multiple sources of noise such as mechanical vibrations of the window-reticle, thermal expansion of the fiber holder and wavelength variations of the laser diode. By analyzing a reconstruction of a hologram sequence of the window-reticle, it is easy to estimate both the reproducibility and the accuracy of the determination of the 3D coordinates of a given object during the experiment.

First, knowing from equation [2.1] that the equivalent distance particle-camera in the magnified reconstructed space is given by $z_{eq} = \frac{z_s z_e}{z_s - z_e}$, it is easy to see that $\frac{\partial z_{eq}}{\partial z_e} = K^2$. Consequently, a couple of particles separated from a small distance Δz_e along the optical axis give rise to a distance $\Delta z_{eq} = K^2 \Delta z_e$ in the reconstruction volume. This is advantageous for an accurate measurement of the axial coordinate because an estimation error of the axial coordinate Δz_r in the image space is reduced to an error $\Delta z_r / K^2$ on the object space. This benefit is observed by the example of reconstruction of the calibrated disks presented in Figure 2.4 where the magnification $K = 3.195 (K^2 \simeq 10)$. As we can see in Figure 2.5, the z-location is estimated with a quite good reproducibility. The measurements have been carried out during the experiment in the wind tunnel over a longer period with a sequence of 2,000 holograms (500 s). The results also show that the set of five disks seems to be submitted together to the same axial variations. This proves the reliability of the best focus plane measurement method. However, the amplitude variations observed on the graph shows the experimental limits of the method: it signifies that the distance of a given particle from the window plane cannot be estimated with an accuracy better than $\Delta z = 20\ \mu m$ during a sequence of several minutes (525 measurements). The estimation of the (x, y) location reproducibility has also been tested on these five calibrated disks. The shift measured from the (x, y) mean position of each disk does not exceed $0.2\ \mu m$ (see [ALL 12]). This result guarantees that the magnitude of a velocity vector can be estimated with an error better than 2% if the time interval between the pulses ensures a displacement larger than 10 pixels, even in the close vicinity of the wall.

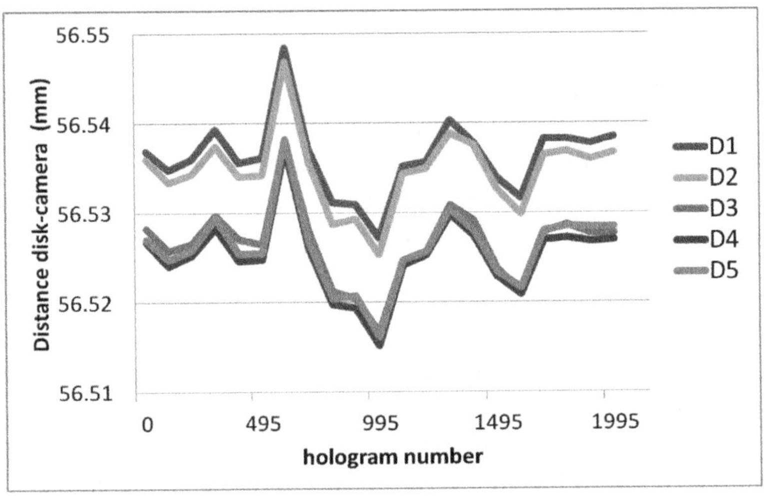

Figure 2.5. *Measured axial location of the five calibrated disk (diameter 5 μm) versus hologram number from a recorded sequence of 500 s*

2.1.3.3. *Velocity measurements*

Three external velocities of a turbulent boundary layer in the large wind tunnel at the Lille Mechanics Laboratory have been tested (i.e. $U_\infty = 3\ m/s, U_\infty = 5\ m/s$ and $U_\infty = 10\ m/s$. About 13,000 holograms have been recorded under different configurations. For illustration, a 3D representation of the 3D velocity field obtained from a sequence of 200 holograms is presented in Figure 2.6. Figure 2.7 shows the measured u component versus the measured distance from the wall Y.

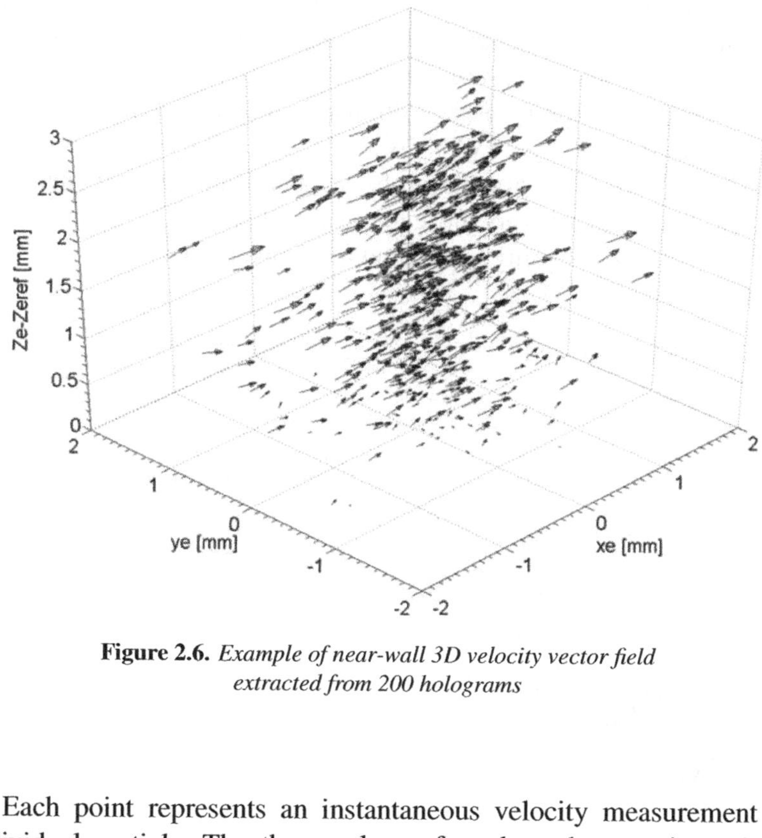

Figure 2.6. *Example of near-wall 3D velocity vector field extracted from 200 holograms*

Each point represents an instantaneous velocity measurement of individual particle. The three values of u_τ have been estimated for particles located near the wall (i.e. small values of Y). Indeed, in this region, as shown by the following equation, the slope of the curve $\bar{u}(Y)$ proportional to the square of the friction velocity:

$$\overline{u} = \frac{u_\tau^2}{\nu} Y \qquad [2.4]$$

where ν is the viscosity of the air ($15.10^{-6}\ m^2/s$). According to this law, we have measured $u_\tau = 0.14 \pm 0.01\ m/s$ for $U_\infty = 3\ m/s$, $u_\tau = 0.21 \pm 0.01\ m/s$ for $U_\infty = 5\ m/s$ and $u_\tau = 0.41 \pm 0.02\ m/s$ for $U_\infty = 10\ m/s$. These results are in good agreement with the characteristics of the boundary layer under investigation [ATK 11]. For an easier comparison with the literature, the variables are plotted in their dimensionless form (i.e. $u^+ = \frac{u}{u_\tau}$ and $y^+ = \frac{u_\tau}{\nu}Y$). Note that the experimental results are mainly contained within the limits given by hot wire anemometry measurements. Also note that the general shape of the velocity measurements is well distributed around the theoretical law in a viscous sublayer [VAN 56]. This preliminary analysis shows the capability of a holographic probe to obtain reliable instantaneous velocity measurements very close to a wall, whose analysis in the field of fluid mechanics is going far beyond the scope of this book.

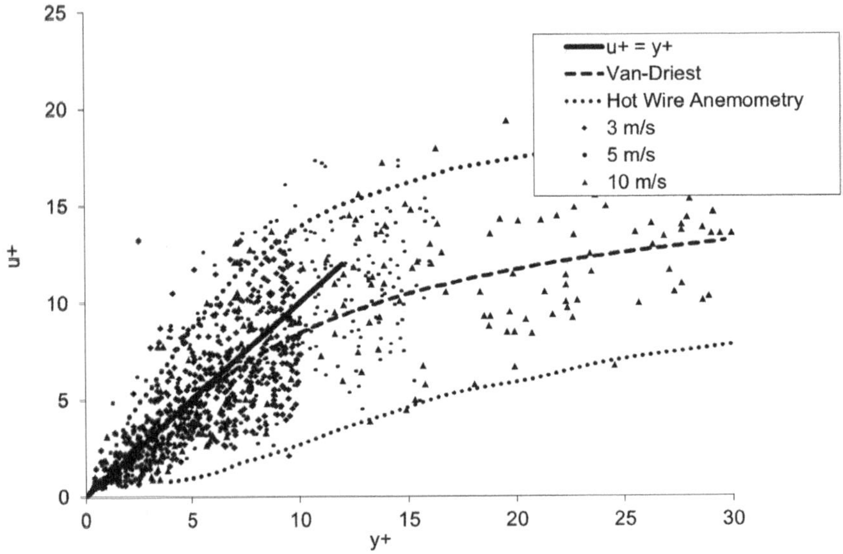

Figure 2.7. *Velocity measurement versus the distance from the wall for three studied velocities $U_\infty = 3{,}5$ and 10 m/s*

2.1.4. *Cavitation bubbles measurements*

The knowledge of the nuclei content is essential in naval hydrodynamics for cavitation inception prediction on propellers and hydrofoils. As a matter of fact, tip vortex cavitation, which is generally the first cavitation to occur on foils or propellers, is very sensitive to the nuclei content. Typically, the nuclei that we are looking at are tiny bubbles whose sizes vary from 20 to 100 μm. Two different types of techniques are generally used: optical techniques that measure the nuclei image size and Venturi techniques that measure the critical pressure of the nuclei to deduce a size distribution of bubbles. Here, DIH has been implemented in the French Large Cavitation Tunnel, the so-called GTH, and comparative measurements using Venturi technique can be found in [LEB 11].

2.1.4.1. *Experimental setup*

As for the previous application, the sample volume is illuminated by a diverging beam coming from a single-mode optical fiber with a 5 μm core. The source is a modulated laser diode emitting at $\lambda = 635\ nm$. The image sensor is a $1,024 \times 1,024$ CMOS camera with 6.7 μm pixels. The curvature radius (taking into account that the wave passes through media of different indices) is set to $z_s = 79.7\ mm$, leading from equation [2.3] to an optimal magnification $K = 1.9$. Note that a direct measurement in the hydrodynamic test section (cross-section of $2\ m \times 1,35\ m$) would have been difficult. Then, an optical probe has been specially designed for these tests. The flow sampling hydraulic loop includes an optical pipe which has an internal square section in order to have two planar optical perspex windows (see Figure 2.8). With this experimental setup, more than 5,000 holograms have been recorded under four different flow pressures ($P = 265\ mb$, $P = 567\ mb$, $P = 927\ mb$ and $P = 1,287\ mb$).

2.1.4.2. *Bubble measurements*

From the best focus reconstructed plane, the particle size is simply determined by counting the pixels whose gray-levels are below a 50% threshold [PU 05]. After having estimated the diameter d_{eq}, the magnification factor K (that depends on the depth coordinate z_{eq}) is applied to estimate the actual particle diameter. The sizing method has been calibrated by using a particle size standard composed of 27 opaque disks deposited by a microlithography technique on a quartz substrate in the range $[2\ \mu m, 1,000\ \mu m]$. Figure 2.9 shows that the measured diameters are in good agreement with the simulations in the upper range.

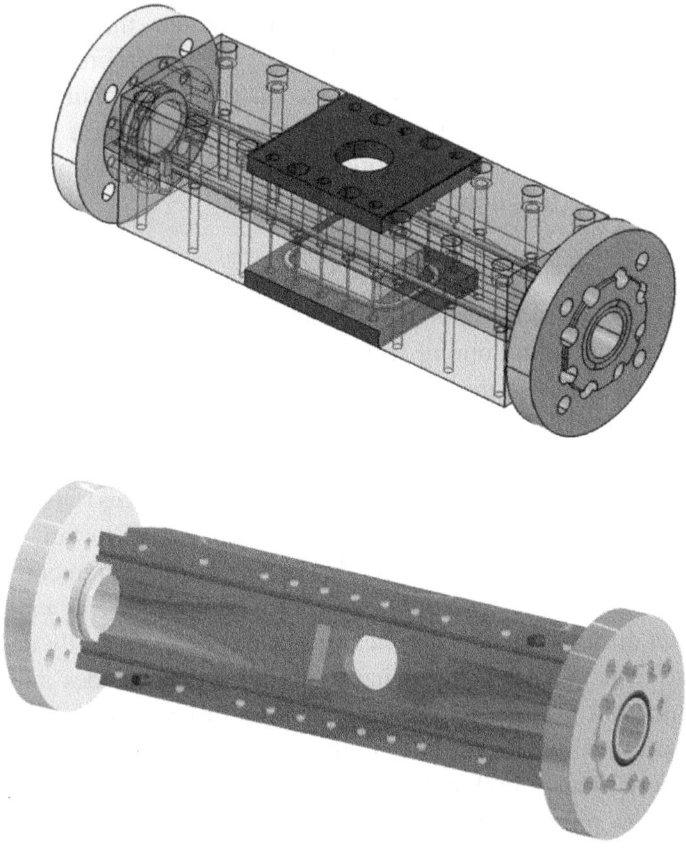

Figure 2.8. *Schematic views of the optical pipe used for recording digital holograms. Holograms are recorded through a circular window of diameter 30 mm*

For diameters below 30 μm, a correction factor must be used, and for the upper size-range [$40 - 200\ \mu m$], we have shown that the bubble diameter d_{eq} can be measured with an accuracy better than 10% [LEB 11].

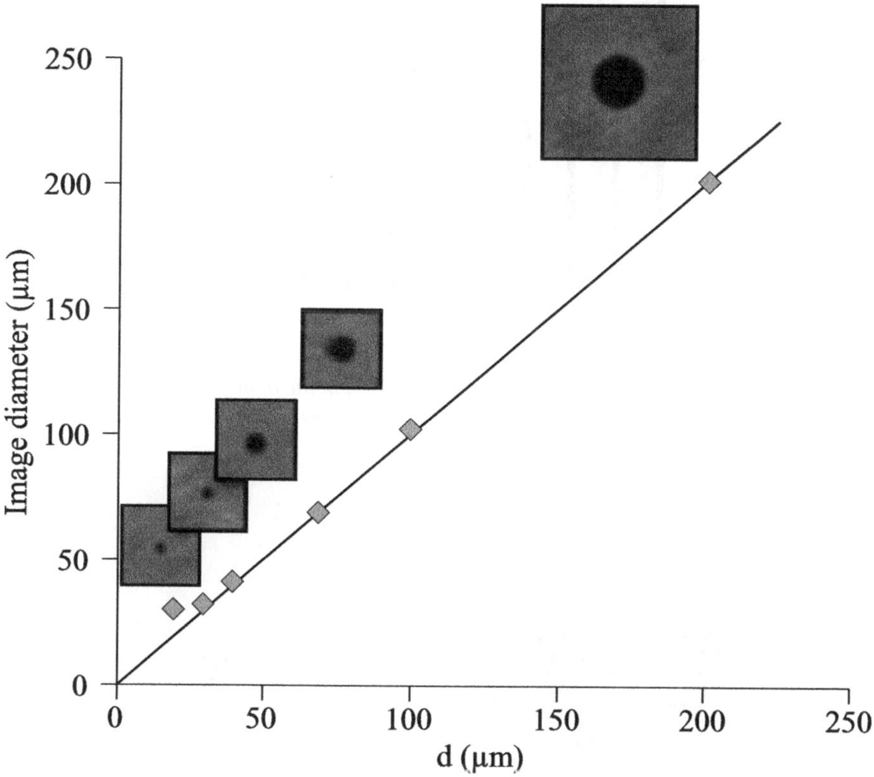

Figure 2.9. *Calibration curve obtained with opaque disks deposited by a microlithography technique on a quartz substrate*

2.1.4.3. *Experimental results*

After having recorded holograms *in situ* (DGA *Techniques Hydrodynamiques*, formerly *Bassin d'Essai des Carènes*), the postprocessing is made possible at CORIA laboratory. Here, 5,000 holograms have been processed by a fully automatic process.

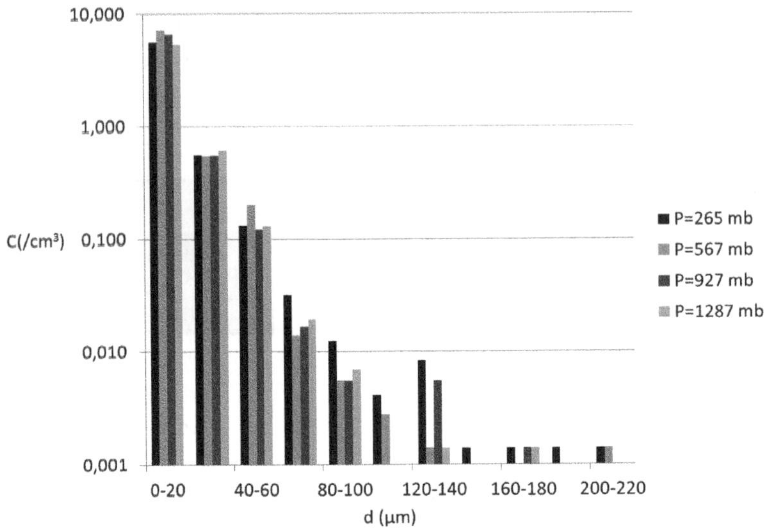

Figure 2.10. *Size nuclei distribution in the optical pipe for different flow pressures*

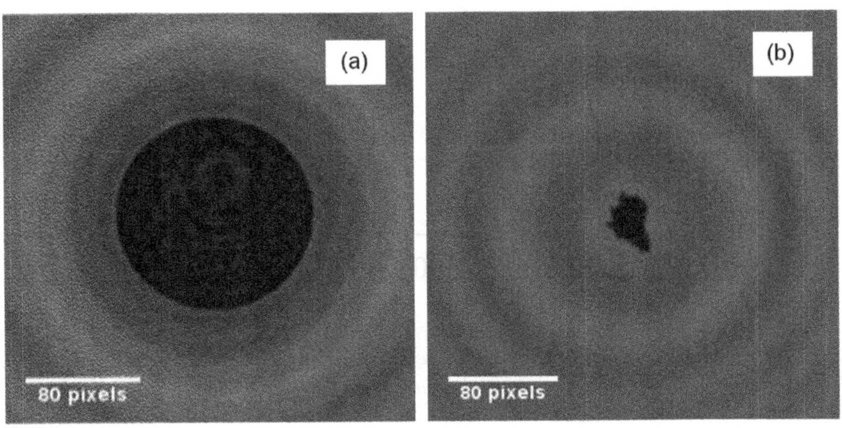

Figure 2.11. *Examples of reconstructed holograms a) Bubble, b) Other particle*

Note that the reliability of diameter measurement is strongly dependent on the capability of depth-coordinate estimation. As described in [MAL 03], the 3D invariance of the point spread function (PSF) is only applicable when the particle image is well focused.

However, it has been shown in [PU 05] that the focusing problem is easy to solve in the range [5–25 μm]. In that case, the depth coordinate is estimated with accuracy better than 100 μm. However, it must be pointed out that for this range, the width of the PSF is roughly $2pixels$, and both non-spherical and spherical particles are seen as circular shape images. So, it is highly probable that other particles such as solid particles could have been wrongly identified as bubbles. This may lead to an overestimation of the concentration of small bubbles. However, for higher diameters, the results presented in Figure 2.10 are in good agreement with the conclusions of [MÉÈ 10]. As shown in Figure 2.11, in the case of larger particles, bubbles can easily be separated from other particles by applying shape criteria.

2.2. The fractional-order Fourier transform

The main objective of this section is to show the elements necessary to understand fractional Fourier transform concisely and precisely. Two notions appear in our title: (1) Fourier transform and (2) fractional order. We do not need to indicate the major implications of the first subject in physics, more specifically its implication on Fourier physical optics and signal processing. The fractional order has nothing to do with fractional calculations. But, for the time being, this underlines a fundamental question, that is whether the description domain of a signal exists between the spatial and spectral domains. Fractional Fourier transform is a generalization of Fourier transform. Some interpretations [OZA 93, LOH 93] and applications in optics have been developed. For example, the authors of [OZA 95, OZA 94, PEL 99] used it to describe the propagation of the light and to analyze spherical mirror resonators. Its mathematical definition is as follows [NAM 80, MCB 87, LOH 93]: the 2D fractional Fourier transform of order a_x for x-cross-section and a_y for y-cross-section with $0 \leq |\alpha_x| \leq \pi/2$ and $0 \leq |\alpha_y| \leq \pi/2$, respectively, of a function $f(x, y)$ is defined as (with $\alpha_p = \frac{a_p \pi}{2}$):

$$\mathcal{F}_{\alpha_x,\alpha_y}[f](x_a, y_a) = \int_{\mathbb{R}^2} N_{\alpha_x}(x, x_a)\, N_{\alpha_y}(y, y_a) f(x, y)\, dx\, dy. \qquad [2.5]$$

The kernel of the fractional operator is defined by:

$$N_{\alpha_p}(x, x_a) = C(\alpha_p) \exp\left(i\pi \frac{x^2 + x_a^2}{s^2 \tan \alpha_p}\right) \exp\left(-\frac{i2\pi x_a x}{s^2 \sin \alpha_p}\right). \quad [2.6]$$

The energy-conservation law is ensured by the coefficient:

$$C(\alpha_p) = \frac{\exp(-i(\frac{\pi}{4}\mathrm{sign}(\sin \alpha_p) - \frac{\alpha_p}{2}))}{|s^2 \sin \alpha_p|^{1/2}} \quad [2.7]$$

which is a function of the fractional order. Here $p = x, y$. The parameter s_p is linked to the number of pixels, denoted N_p, along one transverse axis and to the sampling period, i.e. the pixels size of the CCD sensor, denoted here δ_p by the relation $s_p = N_p \cdot \delta_p^2$. Note that $\mathcal{F}_{\pi/2,\pi/2}$ is the ordinary Fourier transform and $\mathcal{F}_{0,0}$ is the identity operator.

2.3. Digital in-line holography with a sub-picosecond laser beam

DIH has been widely used to visualize flows seeded with small particles [VIK 92, STA 84, THO 67]. The 3D location and size of the particles are determined from the reconstructed images. In many studies of DIH, the incident beam which illuminates the particles is a homogeneous, continuous and collimated plane wave. These coherent sources are sensitive to defects in the optical system, on the one hand, and generate unwanted high spatial frequencies in the plane of the CCD sensor, on the other hand. It is well known that the interaction of such beams with the periodic structures of the pixels matrix creates Moiré effects. Sometimes the optical components (e.g. the lenses used to produce a collimated laser beam or the glass which protects the CCD) create multiple wave interference fringes on the recording plane [BRU 12a, DUB 08]. Some constraints are possible to eliminate the noise sources [SLI 84, KRE 05], in particular to use two types of sources: (1) lasers emitting femtosecond pulses to enlarge the frequency spectrum (partial temporal coherence) and (2) partially spatially coherent source such as the DEL [MAN 95, EYY 07]. Due to their low coherence length, a reduction of parasitic fringes can be expected to reduce the optical noise. In opposition to these drawbacks, a cloud of water droplets or a rotating diffuser can break down the coherence of a coherent source. And if we can analyze these types of optical signals from the coherence point of view, perhaps it is possible to retrieve some characteristics of the noises. We will now study the DIH carried out with such laser sources. The knowledge of the theoretical model in DIH allows us to digitally refocus, in an optical sense, on the objects. Analytical solutions to the problem of scalar diffraction of a partial temporal coherent sub-picosecond pulses and a partially spatially coherent beam by an opaque disk are then given here. In terms of reconstruction, the holographic images can be reconstructed by means of the popular mathematical tool: the 2D fractional Fourier transform [COË 02, PEL 99, NIC 05, REM 13]. Let $\mathcal{O}(\xi, \eta)$ be the transmittance function of the object as shown in Figure 2.12.

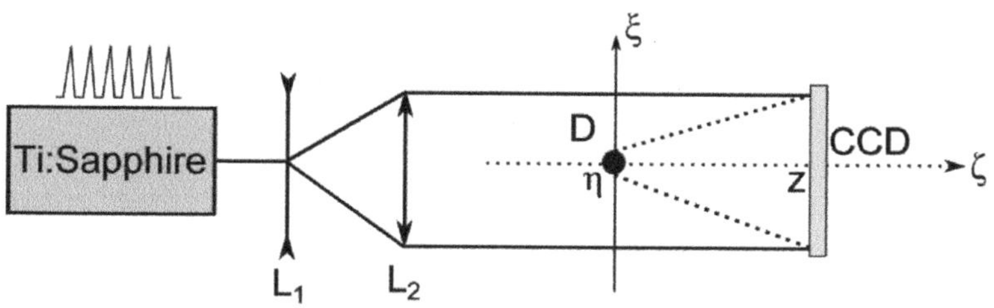

Figure 2.12. *Optical setup*

First, if we consider that this object is illuminated by a monochromatic electromagnetic wave that propagates through a linear, non-dispersive, isotropic and homogeneous medium, then by means of the Huygens–Fresnel diffraction formula, it is given by the scalar integral [GOO 00]:

$$A_z(\mathbf{r}; \omega) = \int_{\mathbb{R}^2} \mathcal{O}(\xi, \eta)\, K_z(\xi, \eta, \mathbf{r}; \omega)\, d\xi\, d\eta \,, \qquad [2.8]$$

where $\mathbf{r} = (x, y)$, $|\mathbf{r}| = r = (x^2 + y^2)^{1/2}$ and

$$K_z(\xi, \eta, \mathbf{r}; \omega) = \frac{\omega}{i2\pi zc} \exp\left[i\frac{\omega}{2cz}((x-\xi)^2 + (y-\eta)^2)\right] \exp\left[i\frac{\omega}{c}z\right]. \quad [2.9]$$

Let us consider a transmittance function $T(\xi, \eta) = \mathbf{1}_S$ with $S = \{(\xi, \eta) | \, \xi^2 + \eta^2 \leq D^2/4\}$, such that the transmission function $\mathcal{O}(\xi, \eta) = 1 - T(\xi, \eta)$ corresponds to the function of an opaque disk of diameter D centered at the origin in the object plane. The complex amplitude $A_z(\mathbf{r}; \omega)$, in the far-field approximation, i.e. $\omega D^2/4zc \ll 1$, in the plane of the CCD sensor is given by:

$$A_z(\mathbf{r}; \omega) = \exp\left[i\frac{\omega}{c}z\right]\left(1 - \frac{\omega}{i2\pi zc}\exp\left[i\frac{\omega r^2}{2zc}\right] F(\mathbf{r}; \omega)\right), \quad [2.10]$$

with

$$F(\mathbf{r}; \omega) = \frac{\pi D^2}{2} \frac{J_1\left[\frac{\omega D}{2zc} r\right]}{\frac{\omega D}{2zc} r}, \quad [2.11]$$

where $\omega = \frac{2\pi c}{\lambda}$ is the frequency of the incident illumination, λ is the wavelength and $c = 3 \cdot 10^8 m/s$ is the velocity of light. The function $F(\mathbf{r}; \omega)$ is the classical Fourier transform of the function T. The intensity distribution in the plane of the CCD sensor, in the far-field approximation [GOO 00, BUR 00], is:

$$I_z(\mathbf{r}) = 1 - \frac{\omega_0}{\pi zc} F(\mathbf{r}; \omega_0) \sin\left[\frac{\omega_0 r^2}{2zc}\right]. \quad [2.12]$$

Under the illumination of an ultrashort laser (e.g. femtosecond laser), the light field is not monochromatic but includes a spectral distribution around a central frequency. Let $u(t)$ be the amplitude of the incident pulsed beam. As the medium between the object and the quadratic sensor is linear, isotropic and homogeneous, the effective frequency range of the Fourier spectrum is invariant. Thus, the propagation of light can be described by a temporal linear filter such as [BON 78a, BON 78b]:

$$E_z(\mathbf{r};t) = [u(t) * A_z(\mathbf{r};t)]_{(t)} \cdot \qquad [2.13]$$

where $E_z(\mathbf{r};t)$ is the diffracted field at time t of an opaque disk illuminated by a pulsed field $u(t)$ and observed at a distance z. The operator $[\cdot * \cdot]_{(t)}$ denotes the temporal convolution. Then we can write the convolution product in the spectral domain as:

$$E_z(\mathbf{r};\omega) = A_z(\mathbf{r};\omega)U(\omega). \qquad [2.14]$$

The quadratic sensor records the temporal average of the light intensity distribution averaged over the time interval $2T$ needed to make an observation. In practice, the disturbance exists only during a finite time $-T \leq t \leq T$, but this interval is a rule so large compared to the physically significant time scales (the mean period $\frac{1}{\nu}$ and the coherence time $\frac{1}{\Delta\nu}$) that we may idealize the situation by assuming $T \longrightarrow \infty$. The expression of the intensity is [BOR 99]:

$$\begin{aligned} I_z(\mathbf{r}) &= \frac{1}{2T}\int_{-T}^{T} E_z(\mathbf{r};t)E_z^*(\mathbf{r};t)\,dt \\ &= \frac{1}{2T}\int_{-\infty}^{+\infty} E_z(\mathbf{r};t)E_z^*(\mathbf{r};t)\,dt. \end{aligned} \qquad [2.15]$$

According to the previous equation and by using the Parseval theorem, we can write that:

$$I_z(\mathbf{r}) = C\int_{-\infty}^{+\infty} \mid A_z(\mathbf{r};\omega) \mid^2 \mid U(\omega) \mid^2 d\omega, \qquad [2.16]$$

where C is a normalized coefficient [GU 96]. C is obtained from the assumption that the on-axis diffraction intensity is unity ($I_z(0) = 1$). If a monochromatic continuous wave (CW) is used, then $U(\omega) = \delta(\omega)$, where $\delta(\omega)$ is the Dirac delta function. Using equations [2.10] and [2.16], it is possible to predict the results of the diffraction pattern produced by an opaque particle illuminated by a short pulsed beam.

We consider here a simple case where the expression of the temporal function of the pulse $u(t)$ is Gaussian such as:

$$u(t) = \exp\left[-i\omega_0 t - \frac{t^2}{T^2}\right], \qquad [2.17]$$

and

$$U(\omega) = T\sqrt{\pi}\exp\left[-\frac{T^2}{4}(\omega - \omega_0)^2\right], \qquad [2.18]$$

its spectrum expression. In the above equation, the variable T represents the time when the amplitude drops to 1/e of the peak amplitude and ω_0 is the central frequency of the incident pulse. The mathematical development is made possible by inserting relations [2.10] and [2.18] into [2.16], and by using the steepest descent method, the analytical solution of the intensity distribution in the far-field approximation recorded by the camera is given by:

$$I_z(\mathbf{r}) \propto 1 - \frac{\omega_0}{\pi z c} F(\mathbf{r}; \omega_0) \sin\left[\frac{\omega_0 r^2}{2zc}\right] \exp\left[-\frac{r^4}{r_0^4}\right], \qquad [2.19]$$

with $r_0 = 2^{1/4}\sqrt{2zcT}$. In equation [2.19], a super-Gaussian function appears, and this term is a function of spatial coordinates (x, y), z and T . By comparing this result with the case of a monochromatic beam, the pulsed laser beam introduces a spatial filter over the intensity distribution by a super-Gaussian. If we take the example of a pulse width $T = 20$ fs and a central wavelength $\lambda = 800$ nm with an opaque disk diameter $D = 150$ μm located at $z = 90$ mm, Figure 2.13 shows the numerical results for the diffraction patterns and gives us the intensity profiles in two cases: with a continue wave source (a) and with a femtosecond laser source (b) to exhibit the natural spatial filter over the intensity distribution by a super-Gaussian.

We can see in these profiles the effect of the finite duration T over the fringes. The number of fringes has been reduced and, consequently, the Moiré effects do not appear in the diffraction pattern. By means of the previous DIH, the image of the particle can be reconstructed with optimal fractional orders, denoted here by α_o [NIC 05]. To do this, the following conditions must be satisfied:

$$\tan \alpha_o = \frac{2\pi z c}{s^2 \omega_0}. \qquad [2.20]$$

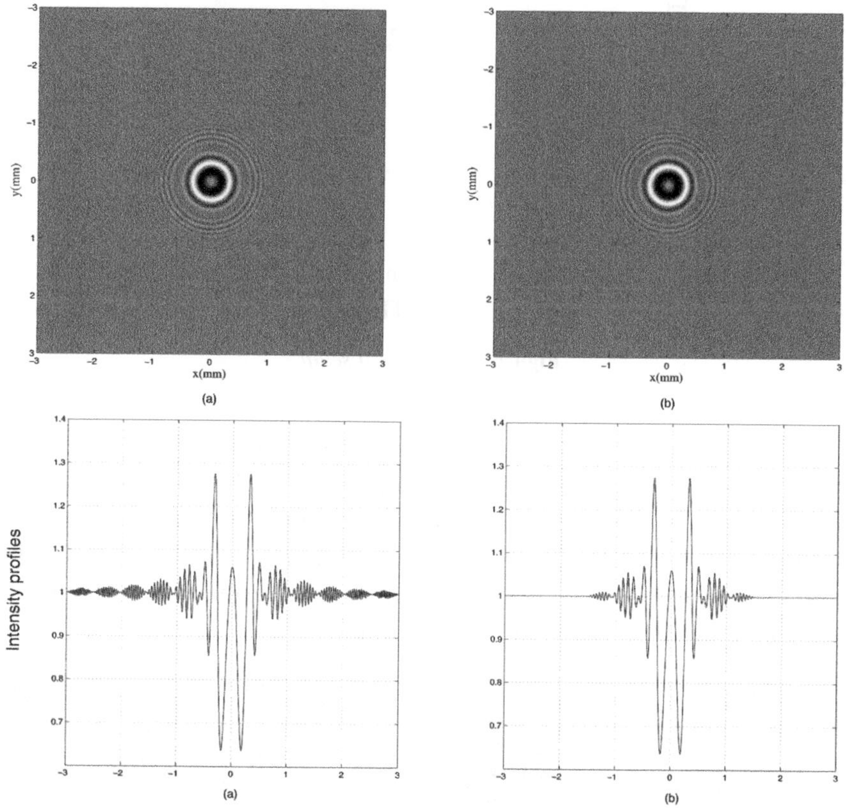

Figure 2.13. *Simulation of the diffraction pattern of an opaque disk with $D = 100\ \mu m$ and located at $z = 80$ mm. Profiles and intensity distributions*

To illustrate the digital reconstruction of the image of the particle, Figure 2.14 represents the process of this reconstruction with the optimal fractional order equal to $\alpha_o = 0.466\pi/2$. This optimal value is obtained from equation [2.20] with $s = 282.8 \cdot \mu$m.

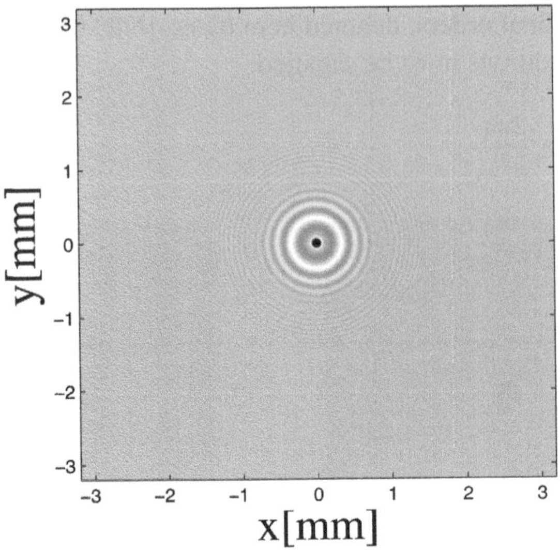

Figure 2.14. *Optical setup*

To illustrate the effects of the femtosecond source under the holograms, we present in Figure 2.15 two cases where the source is a continuous wave in Figure 2.15 on the left and 20 fs-pulsed wave in Figure 2.15 on the right. As we can see, the optical noise which appears by the oblique fringes is eliminated and then this advantage increases the quality of the reconstruction.

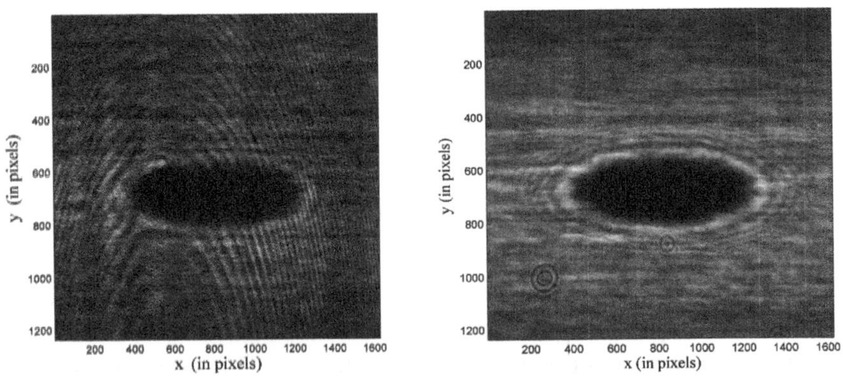

Figure 2.15. *Removing noise by using a femtosecond source. On the left, the CW source and on the right, 20 fs-pulsed laser beam*

The second example is illustrated in Figure 2.16(a). In this case, a femtosecond Ti:sapphire laser at 800 nm with a spectral bandwidth about 14 nm is used. The pulse duration is around 85 fs. The particle diameter of 100 μm is localized at $z = 80\ mm$ from the CCD sensor. Due to the first-order astigmatic aberration of such laser source, it is necessary to consider two different fractional orders along the x-axis and y-axis [NIC 05]. The image of the particle, illustrated in Figure 2.16(b), is reconstructed for the approximate orders $a_{o_x} \simeq 0.473$ and $a_{o_y} \simeq 0.454$. These orders are adjusted in order to reconstruct a focused particle image with a circular shape as the original object $[1 - T]$.

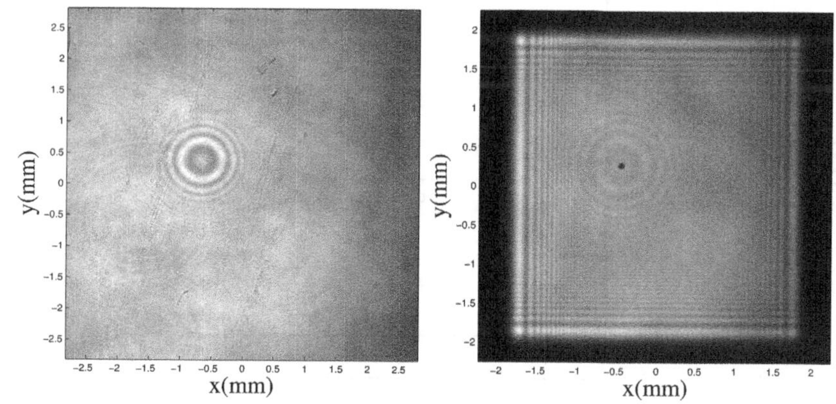

a) Experimental diffraction pattern b) $a_x = 0.473$ and $a_y = 0.454$

Figure 2.16. *Reconstruction of the image of the particle $D = 100\ \mu m$ by means of the fractional Fourier transform*

2.4. Spatially partially coherent source applied to the digital in-line holography

A second possibility to reduce the noises in a digital in-line hologram is to use a light-emitting diode (LED) source. A wide choice of characteristics is possible and they do not affect the following development. Recall that in the case of a monochromatic laser beam, we generally study the propagation of an amplitude by interacting with an optical system and an opaque particle. The main notion we are concerned with here is that in the case of a partially spatially coherent source, we must study the propagation of the mutual intensity function, generally denoted by the letter J. Figure 2.17 describes the context of the study and the methodology used. The characteristic here is an LED at $\lambda_m = 625\ nm$ with an experimental bandwidth in 21.5 nm range. The pinhole localized at the left-hand side of Figure 2.17 is defined by an elliptical Gaussian filter of apertures A_x along the x-axis and A_y along the y-axis (see [SEI 86], equation [29] on p. 786, and [NIC 05] about elliptical astigmatic Gaussian beam).

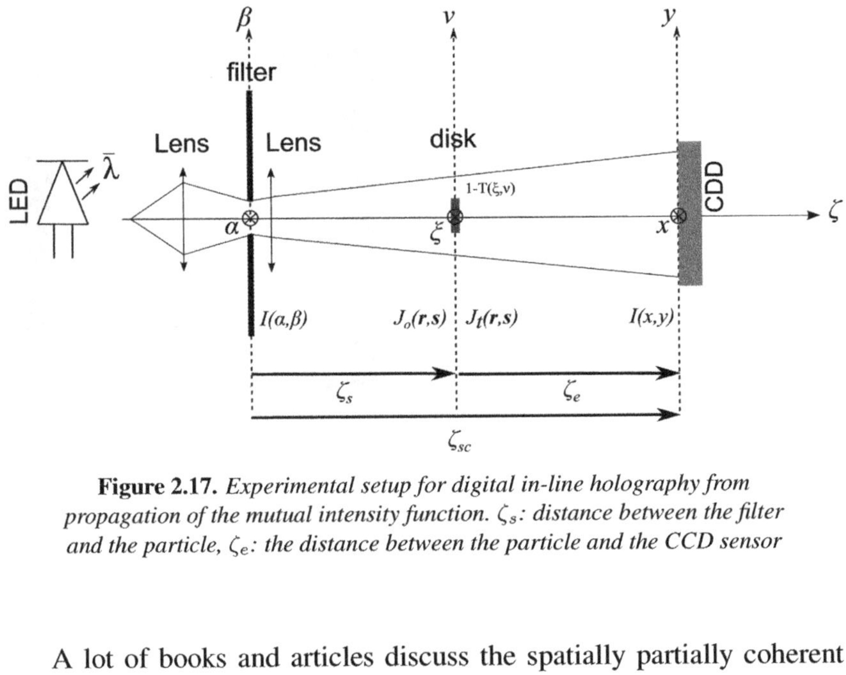

Figure 2.17. *Experimental setup for digital in-line holography from propagation of the mutual intensity function. ζ_s: distance between the filter and the particle, ζ_e: the distance between the particle and the CCD sensor*

A lot of books and articles discuss the spatially partially coherent sources and the probabilistic characteristic of such sources [GOO 00, MAN 95]. Here, we prefer to recall the principal relations from a practical point of view in order to quickly obtain the result about the intensity distribution of the hologram of the opaque particle because the knowledge of the mathematical model allows us to identify the optimal fractional order in the FRFT to reconstruct the image of the particle. Then, from equation [5.6-8] on p. 209 of [GOO 00], the mutual intensity function $\widehat{J_o}$ in the plane just before the opaque disk versus the intensity distribution of the Gaussian filter is expressed by means of Van Cittert–Zernike theorem according to:

$$\widehat{J}_o(\xi, \Delta\xi, \nu, \Delta\nu) = \frac{\lambda_m^2}{\pi(\lambda_m \zeta_s)^2} \exp\left[-i2\pi \frac{\xi\Delta\xi + \nu\Delta\nu}{\lambda_m \zeta_s}\right] \times \int_{\mathbb{R}^2} I(\alpha, \beta) \cdot \exp\left[i \frac{2\pi}{\lambda_m \zeta_s} (\beta\Delta\xi + \alpha\Delta\nu)\right] d\alpha d\beta. \quad [2.21]$$

From [GOO 00], equation [5.7-4] on p. 223, the mutual intensity function, denoted by $\widehat{J}_t$, of the transmitted light is thus:

$$\widehat{J}_t(\xi, \Delta\xi, \nu, \Delta\nu) = \left[1 - T\left(\xi - \frac{\Delta\xi}{2}, \nu - \frac{\Delta\nu}{2}\right)\right] \times \left[1 - \overline{T}\left(\xi + \frac{\Delta\xi}{2}, \nu + \frac{\Delta\nu}{2}\right)\right] \times \widehat{J}_o(\xi, \Delta\xi, \nu, \Delta\nu). \quad [2.22]$$

To calculate the observed intensity distribution of the interferences between the reference beam and the part of this beam diffracted by the opaque disk, we begin with [GOO 00], equation [5.7-6] on p. 224 in the paraxial approximation. Then, the intensity is given by:

$$I(x, y) = \frac{1}{(\lambda_m \zeta_e)^2} \int_{\mathbb{R}^4} \widehat{J}_t(\xi, \Delta\xi, \nu, \Delta\nu) \exp\left[-i \frac{2\pi}{\lambda_m \zeta_e} (\xi\Delta\xi + \nu\Delta\nu)\right] \times \exp\left[i \frac{2\pi}{\lambda_m \zeta_e} (x\Delta\xi + y\Delta\nu)\right] d\xi d\nu d\Delta\xi d\Delta\nu. \quad [2.23]$$

The authors of [COË 11] give the details of the calculus of the intensity distribution. In order to give an illustration, we have first simulated diffraction patterns produced by an opaque disk of 80μm diameter. The full widths at half maximum of the pinhole are $A_x = 50\ \mu$m and $A_y = 50.1\ \mu$m. The particle is localized at $\zeta_s = 216$ mm from the pinhole and at $\zeta_e = 80$ mm from the CCD sensor. Figure 2.18(a) shows the simulated intensity distribution at the observation plane and Figure 2.18(b) shows the experimental intensity distribution recorded by the CCD sensor.

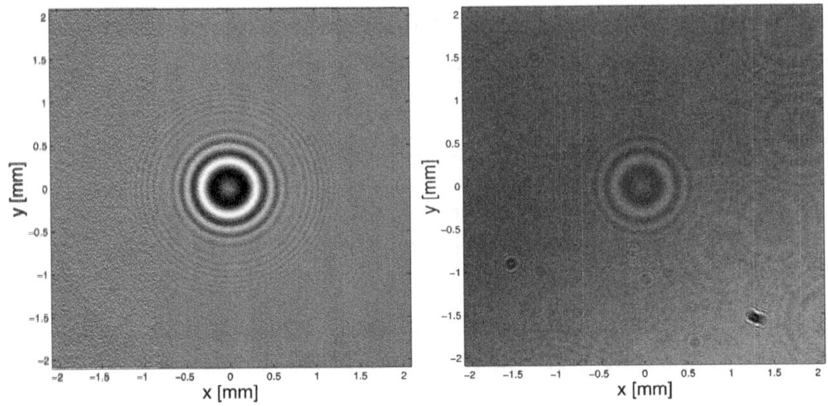

a) Theoretical result b) Experimental result

Figure 2.18. *Intensity distributions for* $A_x = 50\ \mu m$, $A_y = 50.1\ \mu m$, $\zeta_s = 216\ mm$, $D_{th} = 80\ \mu m$, $\zeta_e = 80\ mm$

In the same way as the case of the sub-picosecond pulse, to reconstruct from the FRFT the image of the particle illuminated from a partially spatially coherent beam, the following conditions about the fractional-order Fourier transform

$$\cot\theta_{(x,y)o} = -\frac{s_x^2}{\pi}\cdot\Im\left\{\frac{\pi^2}{(\lambda_m\zeta_e)^2}\frac{\overline{\frac{\pi^2 A^2_{(x,y)}}{(\lambda_m\zeta_s)^2}+i\pi L}}{\left|\frac{\pi^2 A^2_{(x,y)}}{(\lambda_m\zeta_s)^2}+i\pi L\right|^2}\right\},$$

$$L = \left(\frac{1}{\lambda_m\zeta_e}+\frac{1}{\lambda_m\zeta_s}\right) \qquad [2.24]$$

must be satisfied. Here, $\Im\{.\}$ denotes the imaginary part. The coefficients $\theta_{xo,yo}$ are the optimal fractional orders to reconstruct the image of the particle. For example, the reconstructed image of the 80 μm diameter particle from the diffraction pattern in Figure 2.19 with $a_{(x,y)} = 0.833$.

Conversely, the knowledge of these optimized fractional orders leads to the determination of the position ζ_e of the opaque disk from one of the two equations 2.24. Compared to the classical holography where the relation between the fractional order and the axial position of the particle is relatively easy, the relation here is more complicate and it introduces a cubic equation where the axial position, denoted here ζ_e, of the particle is determined by means of the real root of:

$$\zeta_e^3 - a_1 \cdot \zeta_e^2 + a_2 \cdot \zeta_e - a_3 = 0, \qquad [2.25]$$

with the coefficients a_j defined in [COË 11]. Generally, in all optical systems, a magnification must be applied on the reconstructed images to retrieve the real diameter of the object. This point is very important if we want to realize a metrology of the objects. In the case of the DIH with a partially spatially coherent source, the scale factor Δ_j

$$\Delta_j = \frac{\left| \frac{\pi^2 A_j^2}{(\lambda_m \zeta_s)^2} + i\pi L \right|^2}{\pi^2 L^2} \cdot \frac{\lambda_m \zeta_e}{s_j^2 \sin\theta_{jo}}, \qquad [2.26]$$

with $j = x, y$ must be applied over the transverse coordinates.

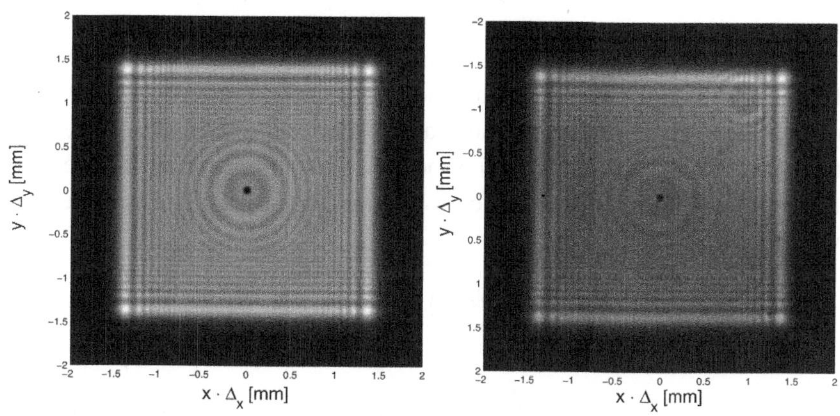

a) From theoretical result, $a_{(x,y)o} = 0.833$ b) From experimental result, $a_{(x,y)o} = 0.834$

Figure 2.19. *Reconstruction of the image of the particle with the FRFT*

2.5. Digital in-line holography for phase objects metrology

2.5.1. *In-line holograms of transparent phase objects*

2.5.1.1. *CW regime*

The analysis of phase objects using digital phase contrast has received much attention since the pioneering work of [CUC 99a, CUC 99b]. Much work has to be done to precisely distinguish opaque objects from purely transparent and, more generally, from partially-transparent phase objects. Recently, we proposed the in-line configuration, with original reconstruction of the phase objects using two-dimensional fractional-order Fourier transforms (2D-FRFT) [COË 02, BRU 09]. Figure 2.20 details the experimental setup that is considered. A Gaussian beam illuminates a phase plate. The expression of the electric field of the beam can be written as:

$$E(\xi,\eta,z) = E_0(z)exp\left(-\frac{\xi^2+\eta^2}{\omega(z)^2}\right)exp\left(-\frac{i\pi}{\lambda}\frac{\xi^2+\eta^2}{R(z)}\right), \qquad [2.27]$$

where ξ and η represent the transverse coordinates of the beam. Furthermore, $\omega(z)$ and $R(z)$ are the width and the radius of curvature of the Gaussian beam along the z-axis. Their expressions are given by the classical formula of Gaussian beams. The amplitude of the electric field that is diffracted at distance z_c behind the phase plate can then be written as:

$$E(x,y,\delta+z_c) = \frac{exp(i\frac{2\pi z_c}{\lambda})}{i\lambda z_c}\int_{-\infty}^{+\infty}\int_{-\infty}^{+\infty}E(\xi,\eta,\delta)$$

$$\times T(\xi,\eta)exp\left(\frac{i\pi}{\lambda z_c}[(\xi-x)^2+(\eta-y)^2]\right)d\xi d\eta \qquad [2.28]$$

where

$$T(\xi,\eta) = \begin{cases} e^{i\varphi} & \text{if } \sqrt{\xi^2+\eta^2}<D/2, \\ 1 & \text{otherwise.} \end{cases} \qquad [2.29]$$

with $\varphi \neq 0$ and D is the diameter of the phase plate. After some mathematical developments, the diffracted field recorded at distance z_c of the phase plate in the plane of the lens is:

$$E(x, y, \delta + z_c) = E_0(\delta) exp\left(i\frac{2\pi z_c}{\lambda}\right)(R + (e^{i\varphi} - 1)O)/(i\lambda z_c)[2.30]$$

with: $R = K^2 e^{\frac{\pi r^2}{\lambda z_c}(iM-N)}$ and $O = \frac{\pi D^2}{2} e^{i\frac{\pi r^2}{\lambda z_c}} T_0(r)$. The system exhibits a symmetry of revolution around the *z*-axis such that we have introduced the radial coordinate: $r = \sqrt{x^2 + y^2}$. The function $T_0(r)$ and all the parameters are detailed in [BRU 09]. Thus, it is possible to exactly simulate the diffraction pattern that will be registered on the CCD camera.

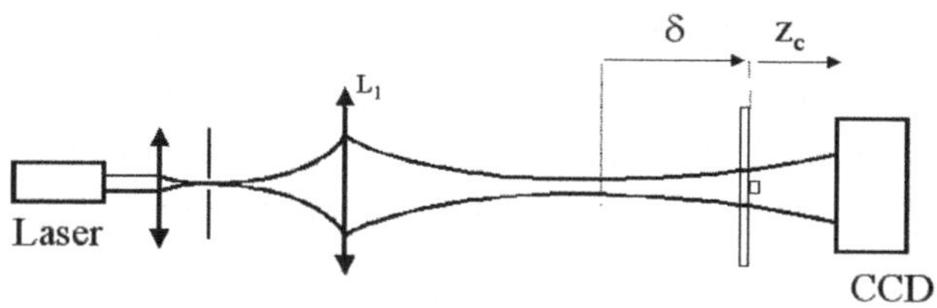

Figure 2.20. *Experimental setup in free space*

2.5.1.2. *General theory for complex setup*

When the phase object is inserted within a complex imaging setup (presence of lenses, windows before and behind the phase object) with a Gaussian beam illumination, it is possible to establish an analytical expression of the electric field in the plane of the CCD sensor. The general setup is described in Figure 2.21. The method consists of combining generalized Huygens–Fresnel integrals whose parameters depend on the coefficients of the optical transfer matrices which describe the system [PAL 97, LAM 98, YUR 87, VER 08], and a description of the phase object through a decomposition of its transmission coefficient over a basis of Gaussian functions [WEN 88].

Then, the expression of the electric field is given by:

$$E(x, y, \ell_1 + \ell_2) = \frac{\omega \exp\left(i\frac{\omega}{c}(n_{\ell_1}\ell_1 + n_{\ell_2}\ell_2)\right)}{i2\pi c\sqrt{B_2^x B_2^y}} \times \left(R(x, y, \ell_1 + \ell_2) + \left(e^{i\varphi} - 1\right) O(x, y, \ell_1 + \ell_2)\right) \quad [2.31]$$

where R is the field that would be obtained in the plane of the CCD sensor in absence of the phase plate, O is the field that would be transmitted by a circular aperture of radius $D/2$, $n_{\ell_1}\ell_1$ is the optical path from the position of the laser to the phase plate and $n_{\ell_2}\ell_2$ is the optical path from the phase plate to the plane of the CCD sensor. These terms are expressed as functions of the different characteristics of the incident beam and the phase plate, and of the coefficients A_1^p, B_1^p, C_1^p, D_1^p, A_2^p, B_2^p, C_2^p, D_2^p (with $p = x$ or $p = y$). They are the coefficients of the optical transfer matrices between the laser and the plate (matrix $M_1^{x,y}$, see Figure 2.21), and the phase object and the CCD sensor (matrix $M_2^{x,y}$, see Figure 2.21). Exact expressions are detailed in [BRU 13]. With this general formalism, cylindrical systems can be described, as the visualization of phase objects within transparent cylindrical pipes or micropipes, by separation of matrices M_1^x and M_1^y, and of matrices M_2^x and M_2^y.

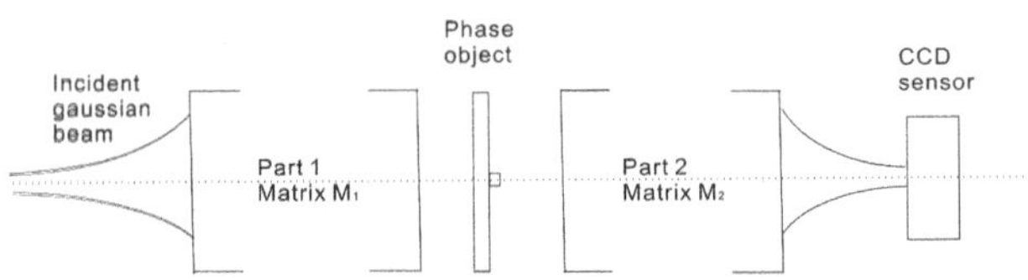

Figure 2.21. *General setup*

2.5.1.3. *Ultrashort pulse illumination*

In the case of an ultrashort beam illumination, the propagation of light is described by a combination of Fresnel diffraction (for each spectral component) and a temporal filter (for proper superposition of monochromatic components) [NIC 06, BRU 12a, BRU 12b]. The CCD sensor records the temporal average of the light intensity distribution. According to Parseval theorem, the diffracted intensity is given by:

$$I(\mathbf{r},\omega) = C\int_0^{+\infty} |\widetilde{E}(\mathbf{r},\omega)|^2 |\widetilde{U}(\omega)|^2 d\omega \qquad [2.32]$$

where C is a constant. The function $\widetilde{U}(\omega)$ is the initial spectrum of the pulses. The Fourier transform of the diffracted field $\widetilde{E}(\mathbf{r},\omega)$ can be evaluated using CW-relations of previous sections. For example, in the most general case of the previous section, $\widetilde{E}(\mathbf{r},\omega)$ can be split into two integrals $\widetilde{R}$ and $\widetilde{O}$, such that:

$$\widetilde{E}\,(x,y,z,\omega) = \frac{\omega \exp\left(i\frac{\omega}{c}\left(n_{\ell_1}(\omega)\ell_1 + n_{\ell_2}(\omega)\ell_2\right)\right)}{i2\pi c\sqrt{B_2^x B_2^y}} \times \left(\widetilde{R}(x,y,z,\omega) + \left(e^{i\varphi(\omega)} - 1\right)\widetilde{O}(x,y,z,\omega)\right) \qquad [2.33]$$

As the spectrum is large, we have to consider carefully that the phase shift introduced by the disk is $\varphi(\omega) = \varphi_o\ \omega/\omega_o$ with $\omega_o = 2\pi c/\lambda_o$, and it depends on ω. λ_o is the central wavelength of the femtosecond laser.

2.5.2. *Reconstruction*

The expression of the diffracted pattern contains a spatially linearly chirped function located around the diffracting object. The principle of reconstruction by 2D-FRFT consists of annihilating this term by multiplication with the complementary quadratic term contained in the 2D-FRFT. In other words, the application of the 2D-FRFT to the pattern acts as a "refocusing" of the image [COË 02]. Depending on the fractional order that is chosen, the longitudinal distance of refocusing can be modified. To reconstruct the image of the particle, the 2D-FRFT of the diffraction pattern is thus numerically evaluated, i.e.:

$$\mathscr{F}_{\alpha_x,\alpha_y}[I] = \mathscr{F}_{\alpha_x,\alpha_y}\left[|R|^2\right] - \mathscr{F}_{\alpha_x,\alpha_y}\left[2Re(R\overline{O}(e^{-i\varphi} - 1))\right] + \mathscr{F}_{\alpha_x,\alpha_y}\left[(2 - 2cos\varphi)|O|^2\right]. \qquad [2.34]$$

The terms $|R|^2$ and $(2 - 2cos\varphi)|O|^2$ do not contain any linear chirp and do not contribute to the reconstruction of the phase object. But the second term on the right-hand side of equation [2.34] contains a linear chirp which is actually present in $R\overline{O}$. As in the case of an opaque object (see [VER 08]), the linear chirp can thus be perfectly analyzed with the 2D-FRFT. It consists of applying the optimal fractional orders that check the following conditions in the case of the setup of Figure 2.21:

$$\alpha_x^{opt} = \arctan\left[\mp\frac{B_2^x\lambda}{s_x^2\left(M_x - D_2^x\right)}\right],$$

$$\alpha_y^{opt} = \arctan\left[\mp\frac{B_2^y\lambda}{s_y^2\left(M_y - D_2^y\right)}\right]. \qquad [2.35]$$

where the different parameters have been defined in [BRU 13] in the case of the setup of Figure 2.21. For example, these general expressions can be applied to the setup of Figure 2.20 using simply $B_2^x = B_2^y = z_c$ and $D_2^x = D_2^y = 1$. It is thus possible to pave the way for the analysis of a phase plate under investigation. The first step consists of applying 2D-FRFT of different orders to the diffraction pattern. When a thin ring due to a high discontinuity is observed, the orders are close to optimal. This gives us the longitudinal position along the z-axis of the phase object through relations [2.35]. In addition, the transverse profile of the reconstructed profile (ring diameter) gives the diameter of the phase object. The magnification factors introduced by the imaging system are given analytically by the parameters of the experimental setup [BRU 12b].

2.5.3. *Experimental results*

The procedure has been tested experimentally. The setup of Figure 2.20 has been realized. The phase plate is an indium tin oxide (ITO) disk of diameter D and width e, which has been deposited on a silica substrate. Other parameters are $\lambda = 632.8\ nm$, $\delta = 12\ cm$, $z_c = 12.5\ cm$, $\varphi = 0.84\pi$ and $D = 280\ \mu m$. Figure 2.22(a) shows the optimal reconstruction of the phase disk. We observe two thin concentric bright and dark circles. The longitudinal position of the phase object is deduced from the optimal order. Figure 2.22(b) shows a radial profile of the reconstructed disk. The limits of the phase objects are clearly visible on both sides where a discontinuity is observed. The distance between both edges of the discontinuity gives the diameter D. The phase shift introduced by the phase disk is obtained from best fitting of the reconstructed profile. Figure 2.22(c) shows two theoretical reconstructed profiles. The thin curve shows the best-fitting profile. It is obtained for $\varphi = 0.9\pi$, while the thick curve is obtained assuming $\varphi = 0.5\pi$. There is approximately a 10% discrepancy between the best-fit phase shift and the actual phase shift $0.84\pi/2$.

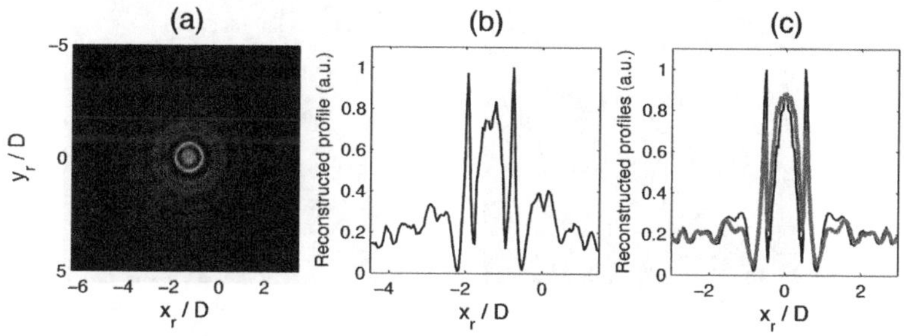

Figure 2.22. *Optimal reconstruction of a circular phase plate a), experimental reconstructed profile b), simulated reconstructed profiles with $\varphi = 0.9\pi$ (thin), and $\varphi = 0.5\pi$ (thick) c). For a color version of this figure, see www.iste.co.uk/picart/digiholography.zip*

In order to present the potentiality of our formalism in the case of anamorphic reconstructions, Figure 2.23(a) shows a hologram predicted when an elliptical phase object is inserted within a transparent cylindrical pipe, while Figure 2.23(b) shows the optimal reconstruction of the elliptical phase object. The form and the size are recovered. The phase shift introduced could be recovered from the analysis of the x-cut and y-cut of the reconstructed profile.

Let us finally illustrate the difference between illuminations in CW regime and femtosecond regime. The setup is the same as given in Figure 2.20. The CCD camera is composed of a CCD sensor and a protection window located just before the sensor (which has no anti-reflection coating in this last case). The window introduces interference fringes due to multiple reflections between its both faces. In CW regime, the patterns recorded with the camera contain additional fringes. The reconstructed pattern is noised (see Figure 2.24(a)). Figure 2.24(b). shows the reconstructed pattern using 20 fs pulses. The sole difference between Figures 2.24(a) and 2.24(b) is the regime of emission of our laser (which can selectively operate in CW or in mode-locked regime). The low coherence length makes possible the elimination of the parasitic fringes [BRU 12a].

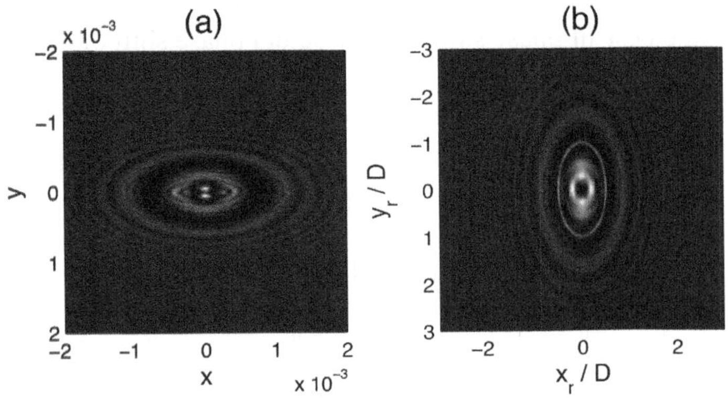

Figure 2.23. *In-line pattern created by an elliptic phase plate inserted in a cylindrical pipe a) and anamorphic reconstruction of the elliptic phase object b)*

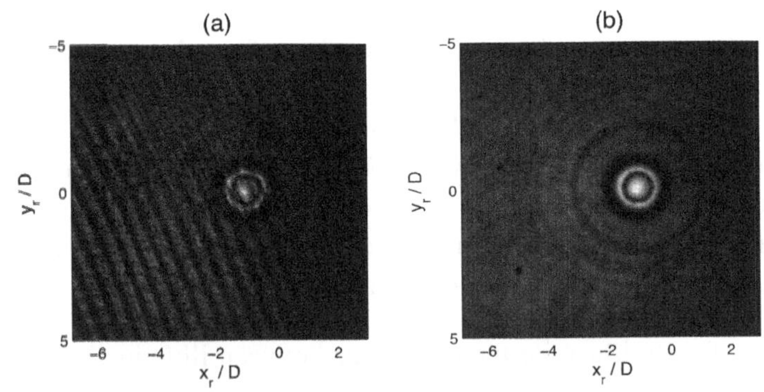

Figure 2.24. *Experimental reconstruction using 2D-FRFT of a circular phase plate in CW regime a), and in femtosecond regime b)*

2.6. Bibliography

[ALL 12] ALLANO D., MALEK M., WALLE F., *et al.*, "Three-dimensional velocity near-wall measurements by digital in-line holography: calibration and results", *Applied Optics*, vol. 52, pp. A9–A17, 2012.

[ATK 11] ATKINSON C., COUDERT S., FOUCAUT J.-M., *et al.*, "The accuracy of tomographic particle image velocimetry for measurements of a turbulent boundary layer", *Experiments in Fluids*, vol. 50, pp. 1031–1056, 2011.

[BON 78a] BONNET G., "Introduction à l'optique métaxiale. Première partie: diffraction métaxiale dans un espace homogène: trilogie structurale, dioptre sphérique", *Annales des Télécommunications*, vol. 33, pp. 143–165, 1978.

[BON 78b] BONNET G., "Introduction à l'optique métaxiale – Deuxième partie : sytèmes dioptriques centrés (non diaphragmés et non aberrants)", *Annales des Télécommunications*, vol. 33, pp. 225–243, 1978.

[BOR 99] BORN M., WOLF E., *Principles of Optics*, Cambridge University Press, 7th ed., 1999.

[BRU 09] BRUNEL M., COËTMELLEC S., LEBRUN D., *et al.*, "Digital phase contrast with the fractional Fourier transform", *Applied Optics*, vol. 48, pp. 579–583, 2009.

[BRU 12a] BRUNEL M., SHEN H., COËTMELLEC S., *et al.*, "Femtosecond digital in-line holography with the fractional Fourier transform: application to phase-contrast metrology," *Applied Physics B: Lasers and Optics*, vol. 106, pp. 583–591, 2012.

[BRU 12b] BRUNEL M., SHEN H., COËTMELLEC S., *et al.*, "Extended ABCD matrix formalism for the description of femtosecond diffraction patterns; application to femtosecond digital in-line holography with anamorphic optical systems", *Applied Optics*, vol. 51, pp. 1137–1148, 2012.

[BRU 13] BRUNEL M., SHEN H., COËTMELLEC S., *et al.*, "Phase contrast metrology using digital in-line holography: general models and reconstruction of phase discontinuities", *Journal of Quantitative Spectroscopy and Radiative Transfer*, vol. 126, pp. 113–121, 2013.

[BUR 00] BURAGA-LEFEBVRE C., COËTMELLEC S., LEBRUN D., *et al.*, "Application of wavelet transform to hologram analysis: three-dimensional location of particles", *Optics and Lasers in Engineering*, vol. 33, pp. 409–421, 2000.

[COË 02] COËTMELLEC S., LEBRUN D.,ÖZKUL C., "Application of the two-dimensional fractional-order Fourier transform to particle field digital holography", *Journal of the Optical Society of America A*, vol. 19, pp. 1537–1546, 2002.

[COË 11] COËTMELLEC S., REMACHA C., BRUNEL M., *et al.*, "Digital in-line holography with a spatially partially coherent beam", *Journal of the European Optical Society – Rapid Publications*, vol. 6, p. 11060, 2011.

[COË 01] COËTMELLEC S., BURAGA-LEFEBVRE C., LEBRUN D., *et al.*, "Application of in-line digital holography to multiple plane velocimetry", *Measurement Science and Technology*, vol. 12, pp. 1392–1397, 2001.

[CUC 99a] CUCHE E., MARQUET P., DEPEURSINGE C., "Simultaneous amplitude-contrast and quantitative phase-contrast microscopy by numerical reconstruction of Fresnel off-axis holograms", *Applied Optics*, vol. 38, pp. 6994–7001, 1999.

[CUC 99b] CUCHE E., BEVILACQUA F., DEPEURSINGE C., "Digital holography for quantitative phase-contrast imaging", *Optics Letters*, vol. 24, pp. 291–293, 1999.

[DUB 08] DUBOIS F., YOURASSOWSKY C., CALLENS N., *et al.*, "Applications of digital holographic microscopes with partially spatial coherence sources," *Journal of Physics: Conference Series*, vol. 139, p. 012027, 2008.

[EYY 07] EYYUBOĞLU H., BAYKAL Y., CAI Y., "Complex degree of coherence for partially coherent general beams in atmospheric turbulence", *Journal of the Optical Society of America A*, vol. 24, pp. 2891–2901, 2007.

[FOU 04] FOURNIER C., DUCOTTET C., FOURNEL T., "Digital holographic particle image velocimetry: 3D velocity field extraction using correlation", *Journal of Flow Visualisation and Image Processing*, vol. 11, pp. 1–20, 2004.

[FUG 04] FUGAL J.P., SHAW R.A., SAW E.W., *et al.*, "Airborne digital holographic system for cloud particle measurements", *Applied Optics*, vol. 43, pp. 5987–5995, 2004.

[GAR 06] GARCIA-SUCERQUIA J., XU W., JERICHO S.K., *et al.*, "Digital in-line holographic microscopy", *Applied Optics*, vol. 45, pp. 836–850, 2006.

[GOO 00] GOODMAN J., *Statistical Optics*, Wiley Classics Library, Wiley, 2000.

[GU 96] GU M., GAN X.S., "Fresnel diffraction by circular and serrated apertures illuminated with an ultrashort pulsed-laser beam", *Journal of the Optical Society of America A*, vol. 13, pp. 771–778, 1996.

[KAT 06] KATZ J., SHENG J., MALKIEL E., "Digital holographic microscope for measuring three-dimensional particle distributions and motions", *Applied Optics*, vol. 45, pp. 3893–3901, 2006.

[KEL 11] KELLY D.P., HEALY J.J., HENNELLY B.M., *et al.*, "Quantifying the 2.5D imaging performance of digital holographic systems", *Journal of the European Optical Society*, vol. 6, p. 11034, 2011.

[KRE 05] KREIS T., *Handbook of Holographic Interferometry: Optical and Digital Methods*, Wiley-VCH, 2005.

[LAM 98] LAMBERT A.J., FRASER D., "Linear systems approach to simulation of optical diffraction", *Applied Optics*, vol. 37, pp. 7933–7939, 1998.

[LEB 11] LEBRUN D., ALLANO D., MÉÈS L., *et al.*, "Size measurement of bubbles in a cavitation tunnel by digital in-line holography", *Applied Optics*, vol. 50, pp. H1–H9, 2011.

[LOH 93] LOHMANN A.W., "Image rotation, Wigner rotation, and the fractional Fourier transform", *Journal of the Optical Society of America A*, vol. 10, pp. 2181–2186, 1993.

[MAL 03] MALEK M., COËTMELLEC S., ALLANO D., *et al.*, "Formulation of in-line holography process by a linear shift invariant system: application to the measurement of fiber diameter", *Optics Communications*, vol. 223, pp. 263–271, 2003.

[MAL 08] MALEK M., LEBRUN D., ALLANO D., "Digital in-line holography system for 3D-3C particle tracking velocimetry", *Applied Physics*, vol. 112, pp. 151–166, 2008.

[MAN 95] MANDEL L., WOLF E., *Optical Coherence and Quantum Optics*, Cambridge University Press, Cambridge, 1995.

[MCB 87] MCBRIDE A., KERR F., "On Namias's fractional Fourier transforms", *IMA Journal of Applied Mathematics*, vol. 39, pp. 159–175, 1987.

[MÉÈ 10] MÉÈS L., LEBRUN D., ALLANO D., *et al.*, "Development of interferometric techniques for nuclei size measurement in cavitation tunnel", *28th Symposium on Naval Hydrodynamics*, Pasadena, Calif., 12–17 September, 2010.

[NAM 80] NAMIAS V., "The fractional order Fourier transform and its application to quantum mechanics", *Journal of the Institute of Mathematics and Its Applications*, vol. 25, pp. 241–265, 1980.

[NIC 05] NICOLAS F., COËTMELLEC S., BRUNEL M., *et al.*, "Application of the fractional Fourier transform to digital holography recorded by an elliptical, astigmatic Gaussian beam", *Journal of the Optical Society of America A*, vol. 22, pp. 2569–2577, 2005.

[NIC 06] NICOLAS F., COËTMELLEC S., BRUNEL M., *et al.*, "Digital in-line holography with a sub-picosecond laser beam, *Optics Communications*, vol. 268, pp. 27–33, 2006.

[OWE 00] OWEN R.B., ZOZULYA A.A., "In-line digital holographic sensor for monitoring and characterizing marine particulates", *Optical Engineering*, vol. 39, pp. 2187–2197, 2000.

[OZA 93] MENDLOVIC D., OZAKTAS H.M., "Fractional Fourier transforms and their optical implementation: I", *Journal of the Optical Society of America A*, vol. 10, pp. 1875–1881, 1993.

[OZA 94] OZAKTAS H.M., MENDLOVIC D., "Fractional Fourier transform as a tool for analyzing beam propagation and spherical mirror resonators," *Optics Letters*, vol. 19, pp. 1678–1680, 1994.

[OZA 95] OZAKTAS H.M., MENDLOVIC D., "Fractional Fourier optics", *Journal of the Optical Society of America A*, vol. 12, pp. 743–751, 1995.

[PAL 97] PALMA C., BAGINI V., "Extension of the Fresnel transform to ABCD systems", *Journal of the Optical Society of America A*, vol. 14, pp. 1774–1779, 1997.

[PAN 01] PAN G., MENG H., "Digital in-line holographic PIV for 3D particulate flow diagnostics", *4th International Symposium on Particle Image Velocimetry*, Göttingen, Germany, 17–19 September 2001.

[PEL 99] PELLAT-FINET P., "Diffraction entre un émetteur et un récepteur localement toriques. Application à l'étude des systèmes astigmates", *Comptes Rendus de l'Académie des Sciences - Series IIB - Mechanics-Physics-Astronomy*, vol. 327, pp. 1269–1274, 1999.

[PIC 08] PICART P., LEVAL J., "General theoretical formulation of image formation in digital Fresnel holography", *Journal of the Optical Society of America A*, vol. 25, pp. 1744–1761, 2008.

[PU 05] PU S.L., ALLANO D., PATTE-ROULAND B., *et al.*, "Particle field characterization by digital in-line holography: 3D location and sizing", *Experiments in Fluids*, vol. 39, pp. 1–9, 2005.

[REM 13] REMACHA C., COËTMELLEC S., BRUNEL M., *et al.*, "Extended wavelet transformation to digital holographic reconstruction: application to the elliptical, astigmatic Gaussian beams", *Applied Optics*, vol. 52, pp. 838–848, 2013.

[SEI 86] SIEGMAN A., *Lasers*, University Science Books, 1986.

[SHE 03] SHENG J., MALKIEL E., KATZ J., "Single beam two-views holographic particle image velocimetry", *Applied Optics*, vol. 42, pp. 235–250, 2003.

[SHE 08] SHENG J., MALKIEL E., KATZ J., "Using digital holographic microscopy for simultaneous measurements of 3D near wall velocity and wall shear stress in a turbulent boundary layer", *Experiments in Fluids*, vol. 45, pp. 1023–1035, 2008.

[SLI 84] SLIMANI F., GREHAN G., GOUESBET G., *et al.*, "Near-field Lorenz-Mie theory and its application to microholography", *Applied Optics*, vol. 23, pp. 4140–4148, 1984.

[STA 84] STANTON A.C., CAULFIELD H.J., STEWART G.W., "An approach for automated analysis of particle holograms", *Optical Engineering*, vol. 5, p. 235577, 1984.

[STA 13] STANISLAS M., ABRANTES J.K., COUDERT S., *et al.*, "Digital microscopic holography for micrometer particles in air", *Applied Optics*, vol. 52, pp. A397–A409, 2013.

[THO 67] THOMPSON B., WARD J., ZINKY W., "Application of hologram techniques for particle size analysis", *Applied Optics*, vol. 6, pp. 519–526, 1967.

[VAN 56] VAN DRIEST E.R., "On turbulent flow near a wall", *Journal of Aeronautical Sciences*, vol. 23, pp. 1007–1011, 1956.

[VER 08] VERRIER N., COËTMELLEC S., BRUNEL M., *et al.*, "Digital in-line holography in thick optical systems: application to visualization in pipes", *Applied Optics*, vol. 47, pp. 4147–4157, 2008.

[VIK 88] VIKRAM C.S., BILLET M.L., "Some salient features of in-line Fraunhofer holography with divergent beams", *Optik*, vol. 78, pp. 80–83, 1988.

[VIK 92] VIKRAM C., *Particle Field Holography*, Cambridge University Press, 1992.

[WEN 88] WEN J.J., BREAZEALE M., "A diffraction beam expressed as the superposition of Gaussian beams", *Journal of the Acoustical Society of America*, vol. 83, pp. 1752–1756, 1988.

[XU 03] XU W., JERICHO M.H., KREUZER H.J., "Tracking particles in four dimensions with in-line holographic microscopy", *Optics Letters*, vol. 28, pp. 164–166, 2003.

[YUR 87] YURA H.T., HANSON S.G., "Optical beam wave propagation through complex optical system", *Journal of the Optical Society of America A*, vol. 4, pp. 1931–1948 1987.

[ZHA 06] ZHANG Y., SHEN G., SCHRÖDER A, *et al.*, "Influence of some recording parameters on digital holographic particle image velocimetry", *Optical Engineering*, vol. 45, p. 075801, 2006.

3

Digital Color Holography For Analyzing Unsteady Wake Flows

The fast development of technology, such as high-resolution sensors, various Diode-Pumped Solid-State (DPSS) lasers with long coherence, data postprocessing and computation power, provides us with the opportunities to develop new optical methods capable of simultaneous full-field measurements with high spatial and temporal resolutions and to retrieve absolute data. Digital holography with matrix sensors appeared in the last decade with cheap high-resolution charge-coupled device (CCD) cameras and the increasing power of computers [SCH 94]. Image sensors now have size and spatial resolutions compatible with the requirements for digital holographic recording. For example, matrices with 1,636 × 1,238 pixels sized 3.9 × 3.9 μm^2 are now available [YAM 97]. In the literature, some papers concerning works in digital color holographic interferometry and they demonstrated phase shifting digital color holography using a multiwavelength HeCd continuous wave laser (636, 537.8 and 441.6 nm) and a color CCD equipped with a Bayer mosaic[1] [YAM 02, KAT 02]. The authors showed the possibility for the reconstruction of color images, but experimental results have a relatively low spatial resolution since the effective pixel number at each wavelength was

Chapter written by Jean Michel DESSE and Pascal PICART.

1 Matrix structure in front of a sensor to create a color information from a panchromatic monochrome.

818 × 619, leading to an effective pixel pitch of 7.8 μm. Demoli *et al.* [DEM 03] presented the first study on fluids using digital color Fourier holography. They used a monochrome CCD sensor and three wavelengths from three continuous wave lasers (647 nm, 532 nm, 476 nm). Their results showed the evolution of thermal dissipation in an oil tanker with an excellent resolution, by using the amplitude image of the reconstructed holograms [DEM 03]. Note that, historically, the use of several wavelengths for holographic interferometry was first described by Jeong *et al.* [JEO 97]. Furthermore, *École Polytechnique Fédérale de Lausanne's* researchers [KUN 07] and INOA [FER 07] proposed the use of several wavelengths in digital holographic microscopy. For quantitative phase microscopy, Ferraro *et al.* [FER 04] showed that severe chromatic aberration can be eliminated. Since the main effect of the chromatic aberration is to shift the correct focal image plane differently at each wavelength, this can be readily compensated by adjusting the corresponding reconstruction distance for each wavelength. In these works, the recording at each wavelength is sequential. Such an approach was applied in three-dimensional (3D) image fusion using sequential color recording at several distances [JAV 05]. This strategy was also utilized with three laser wavelengths as an imaging approach [ZHA 08]. All these methods use sequential recording at each wavelength, where off-axis reference waves are general incident on the recording area at a constant angle. At the same time, two-wavelength profilometry was proposed for a digital holographic microscope using two laser diodes (680 and 760 nm) and a monochrome sensor, giving a synthetic wavelength of 6.428 μm [KUH 07]. The method is based on spatial multiplexing by incoherent addition of single-wavelength interferograms, in the same way as was developed by Picart *et al.* with a single wavelength, each having different propagation directions for the reference waves, and recording with a monochrome CCD [PIC 03, PIC 05].

In regards to these works, ONERA and LAUM[2] joined their competences acquired in the past in order to develop new adaptable

2 *Laboratoire d'Acoustique de l'Université du Maine*, Le Mans, France.

optical imaging methods, firstly having properties such as full-field imaging with high spatial and temporal resolutions, secondly giving absolute data after postprocessing and finally giving dynamic 3D measurements. These non-invasive optical methods are based on digital color holography [DES 08, DES 12]. After presenting the advantages associated with the use of a polychromatic light source rather than monochromatic, digital color holographic interferometry is applied for analyzing three different aerodynamic cases: the unsteady wake flow downstream of a cylinder at a subsonic Mach number, the supersonic jet presenting high-density gradients and a small hydrogen jet in the shock layer at hypersonic Mach number.

The holographic numerical processing of all the methods presented in this chapter is based on off-axis holography and filtering in the Fourier spectrum to extract the complex amplitude of the object wave. Authors invite the reader to have a look at Chapter 1 in order to be familiar with the digital processing (section 1.1.3.1). Note that in many cases, the image of the useful area of interest is projected onto the sensor plane, so that there is no particular requirement for numerical refocusing. It follows that the optical phase can be directly extracted from the complex field and that phase differences can be computed from a sequence of digitally recorded holograms.

3.1. Advantage of using multiple wavelengths

In monochromatic interferometry (e.g. $\lambda = 647$ nm), it is well known that the classical interference pattern is represented by a succession of dark and bright red fringes. For two successive fringes, the optical path difference is equal to the wavelength of the laser source (Figure 3.1(a)). Unfortunately, the zero order of interference fringes can never be identified and it is one of the major difficulties with interference fringes in monochromatic light. Sometimes, it is not possible to follow the displacement of the fringes through a shock wave, for example, or to count the fringe number in a complex flow. When the light source is a continuous source (500 W xenon, see Figure 3.1(b)), the interference pattern is a colored fringe pattern in a sequence approximately matching Newton's color scale. This fringe diagram shows a unique white fringe, visualizing the zero order of interference and measures very small path differences, because six or seven different colors define the interval 0–0.8 µm. However, when the path difference is greater than 3 or 4 µms, the colors can no longer be separated and the larger path differences cannot be correctly measured [DES 97]. Figure 3.1(c) shows the fringes obtained with a laser that emits three different wavelengths (one blue line, one green line and one red line).

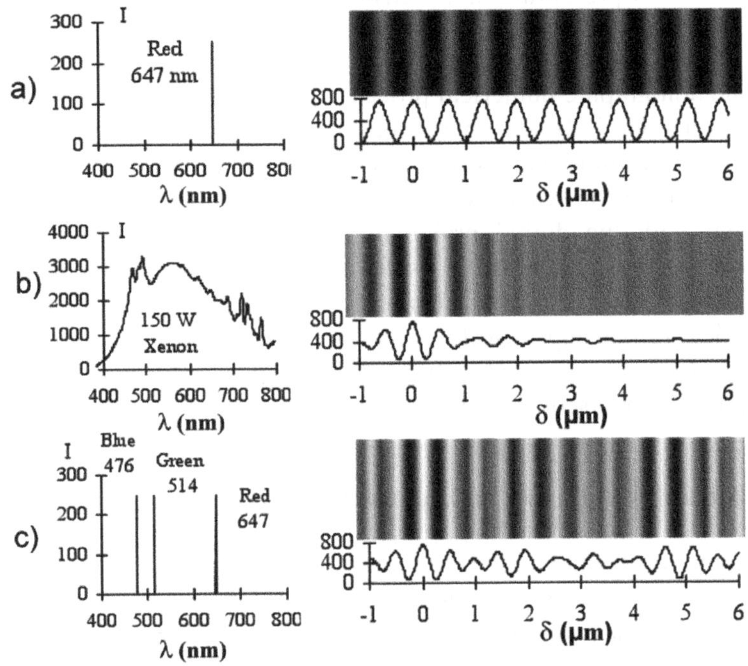

Figure 3.1. *Spectra and interference fringes given by three different light sources. For a color version of this figure, see www.iste.co.uk/picart/digiholography.zip*

We can see that the disadvantages of the two other sources have disappeared. The zero order is always identifiable and the colors always remain distinguishable for the small and the large path differences. The interference pattern also presents the following peculiarity: while the white fringe is not visible on the interferogram, the sequence of three successive colors in the diagram is unique. An example is given in Figure 3.2 to illustrate the color contribution for directly determining the sign of the change in the oil film thickness from the knowledge of the color in the Newton's tints scale [DES 03]. Three different light sources are used to record the oil film interference fringes in the vicinity of the boundary layer transition region on a flat plate. The flow is from the left to right and the central part of the interferogram has been isolated in order to not take into account the test section lateral boundary layers.

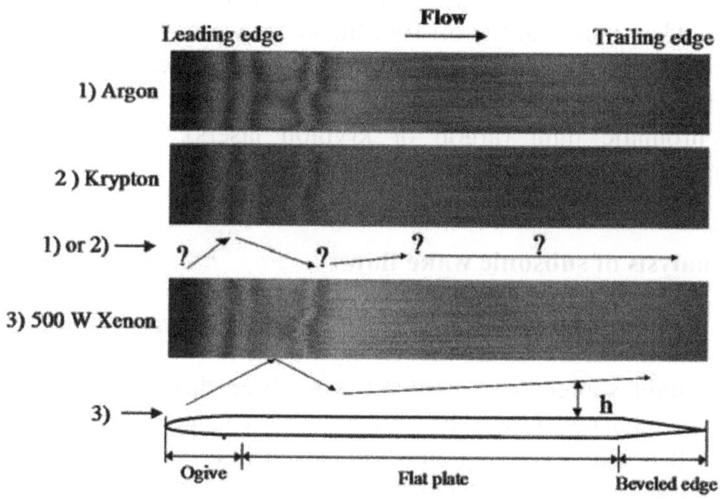

Figure 3.2. *Interference fringes obtained with three different light sources. For a color version of this figure, see www.iste.co.uk/picart/digiholography.zip*

Interferograms 1 and 2 show interference fringes recorded with green and red lines, respectively, issued from the argon/krypton laser. The only visualization of dark and bright fringes does not lead to the unambiguous determination of the evolution of the oil film thickness. Interferogram 3 is obtained with a xenon light source and the knowledge of Newton's tints scale leads to the variation of the oil film thickness without any doubt. Moreover, because no color is enclosed with two identical tints in Newton's scale, it is easy to detect the extrema of the oil film thickness profile. This is important because the gradient changes of the oil film profile are determined very easily. In interferogram 3 of Figure 3.2, we can see that the pale green fringe is enclosed within two fringes having the same red color (first change in the oil film thickness slope sign) and that the yellowish fringe is enclosed within two identically purplish–red colored fringes (second change of the slope sign). The analysis of the interferogram colors indicates that the oil film thickness is increasing upstream of the location of the pale green color, and then decreasing up to that of the yellowish color and increasing again downstream of it. This information is very difficult to obtain with a monochromatic light source. Also, note that the oil film thickness varies little downstream. Better sensitivity is obtained with a white light source because several fringes are visible downstream of the central part of the model. In monochromatic light (argon or krypton lasers), it is possible to distinguish a variation of only one fringe.

3.2. Analysis of subsonic wake flows

3.2.1. *Description of the digital color holographic interferometer*

The light source is constituted with three different DPSS lasers (457, 532 and 660 nm). The assembly shown in Figure 3.3 is very simple. It is like a conventional Michelson interferometer in which a beam splitter cube is inserted between the spatial filter and the test section. The spatial filter is placed at the focal point of the achromatic lens so that the test section is illuminated with a parallel light beam as in previous optical setups. 50% of the light is reflected from the concave spherical mirror to form the three reference beams and 50% of light passes through the test section to form the three measurement waves. The flat mirror, placed just behind the test section, returns the beams toward the beam splitter. Twenty-five percent of light is focused on the diaphragm that is placed in front of the camera lens. So, 25% of the reference beam intensities are focused in the same diaphragm by the concave mirror.

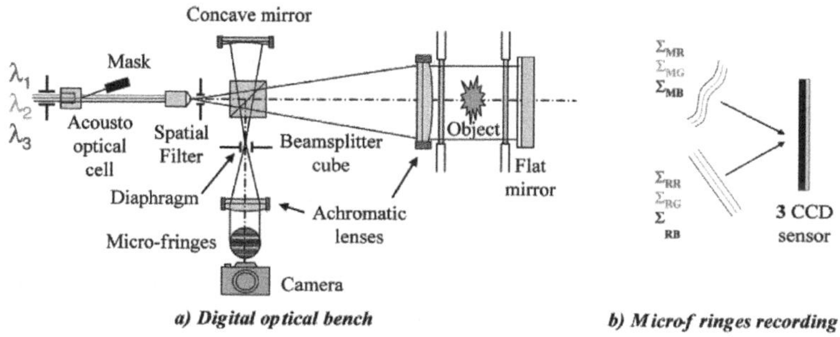

Figure 3.3. *Digital color holographic interferometer – formation of spatial carrier frequencies. For a color version of this figure, see www.iste.co.uk/picart/digiholography.zip*

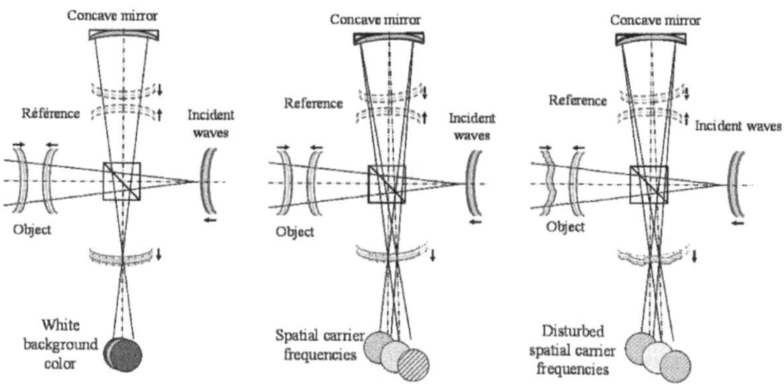

Figure 3.4. *Generation and microfringes formation by the phenomenon studied. For a color version of this figure, see www.iste.co.uk/picart/digiholography.zip*

The generation of microfringes used as spatial carrier frequencies is shown in Figure 3.4. When the focal points of the reference and object waves are superimposed in the diaphragm, which is placed in front of the camera, a uniform background color is observed on the screen. If the focusing point of the reference waves is moved in the plane of the diaphragm, straight interference fringes are introduced into the field of view. These microfringes are recorded on the CCD in order to calculate the three reference phase maps. Then, the wind tunnel is started and the three object waves are distorted by the aerodynamic phenomenon. Microfringe interferences are again recorded by the color sensor to enable calculation of the phase maps related to the object. For phase map differences, the reference phase is subtracted from the phase object.

3.2.2. *Results obtained with subsonic wake flows*

Near wake flow downstream of a circular cylinder was studied at Mach 0.45. An ORCA-3CCD Hamamatsu camera with three chips of 1,344 × 1,024 pixels, whose size is of 6.45 µm × 6.45 µm, was used. The frame rate is 9 im/s and the filters of the RGB camera are very narrow and centered on the three laser wavelengths. As the frame rate is very slow compared to the frequency of the vortex street, a transducer has been implemented in the cylinder at an azimuth of 90° (perpendicular to the flow axis) in order to synchronize the interferogram recording with the signal of the unsteady pressure measurement. The cycle of the vortex street was decomposed in eight different instants shifted by 76 µs, and, at each instant, five interferograms were recorded from several cycles to average the unsteady maps. Figure 3.5 shows two microfringe images recorded with and without the flow in order to constitute reference and object interferograms. It can be seen in the zoomed image that microfringes are deformed by the shear layer of the upper side.

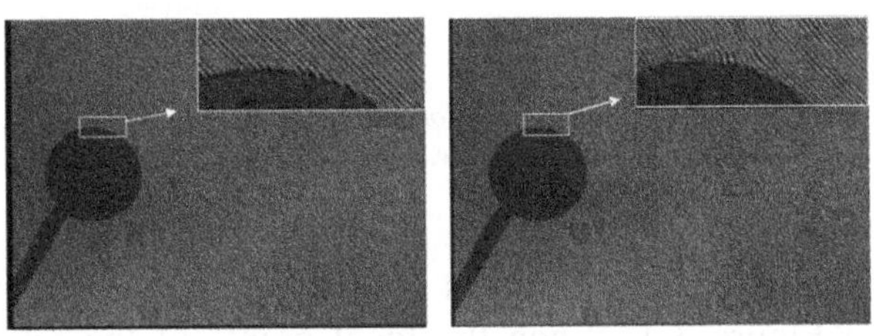

Figure 3.5. *Microfringes recording for the reference and measurement holograms. For a color version of this figure, see www.iste.co.uk/picart/digiholography.zip*

The three Fourier transforms are calculated from each image in order to reconstruct the phase maps with the +1 order (the zero order and the −1 order are filtered). An example of reference and measurement spectra is given in Figure 3.6 for the green line. We can see that the spectrum only shows a spot corresponding to the green spatial carrier frequency. No parasitic frequencies due to the blue and red lines are found. By subtracting the reference phase maps from the measurement phase maps, we obtain the modulo 2π phase difference maps. After unwrapping, the computation of the refractive index maps and the gas density field assuming the Gladstone–Dale relation is possible.

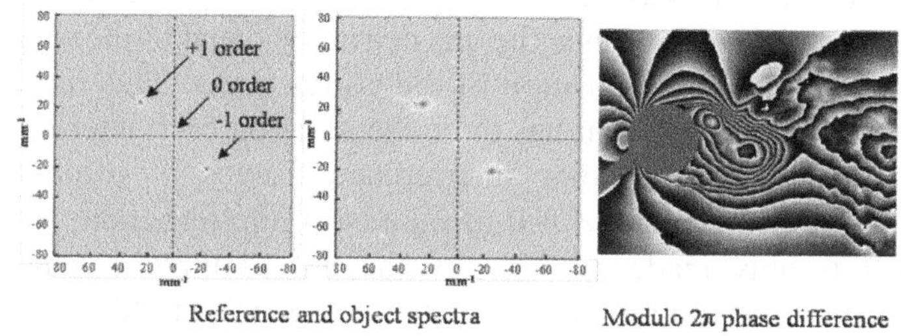

Figure 3.6. *FFT reference and object spectra and phase difference map for the green line. For a color version of this figure, see www.iste.co.uk/picart/digiholography.zip*

Color interference fringes and gas density field are shown in Figure 3.7 for the first three images of one cycle of the vortex street.

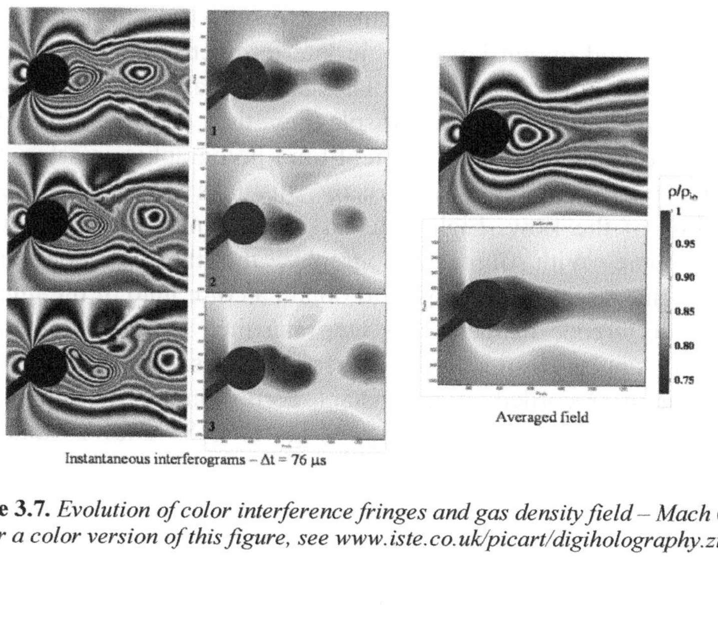

Figure 3.7. *Evolution of color interference fringes and gas density field – Mach 0.45. For a color version of this figure, see www.iste.co.uk/picart/digiholography.zip*

The intensity of the interference fringes is computed on the three channels R, G and B from the phase maps with the following relationship:

$$I_\lambda = A_\lambda(1+\cos(\Delta\varphi_\lambda) \qquad [3.1]$$

where A_λ is the amplitude of the object wave and $\Delta\varphi_\lambda$ is the temporal phase difference at wavelength λ. The gas density measured at the cylinder nose is striking since the gas density is equal to the stagnation gas density through the position of the vortex street. This means that the color found at this point has to be the same on each interferogram. Here, the intensity of color interference fringes is computed by imposing the white color ($\delta = 0$ µm) on each interferogram. Note that this shift is only made possible by the use of three colors in the experiments. The time evolution of the gas density fields shows that the gas density decreases to 73% of ρ_{io} in the vortex core. Then, the averaged field of one cycle is calculated by averaging eight maps of instantaneous gas density field.

3.2.3. *Comparison between holographic plate and digital holograms*

In regards to previous experimental results [DES 09], the silver-halide plate and digital holographic interferometry can be compared. The only possibility to compare plates and digital interferograms is to compare the interferograms displaying the interference fringes. Indeed, the technique based on real-time holographic interferometry using panchromatic plates directly displays the color density variations of the flow. Light intensity is obtained, and not a phase measurement. With digital holography, the three monochromatic intensity maps are superimposed to obtain a color map of the intensity of the interference fringes. This map can then be compared to that obtained using the reflection holographic plates. After extracting an interferogram recorded at an instant corresponding to that of the digital interferogram recording, Figure 3.8 shows that the correspondence is very good.

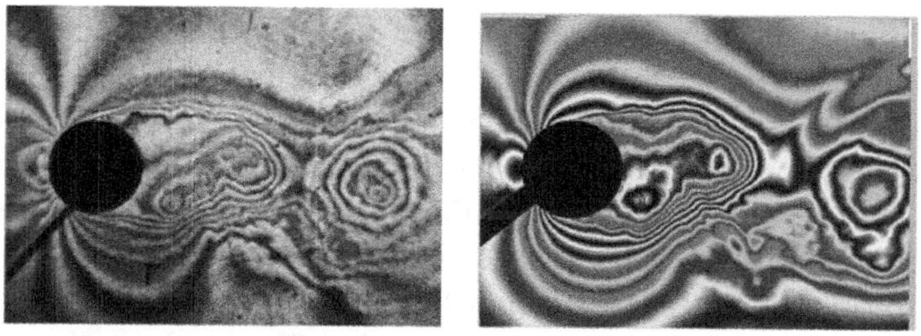

Figure 3.8. *Comparison between real-time and digital interferograms. For a color version of this figure, see www.iste.co.uk/picart/digiholography.zip*

So, with silver-halide color holographic interferometry, a reflection panchromatic holographic plate (7,000–10,000 lines per millimeter in spatial resolution) has to be illuminated with a total energy at 600 µJ and the resetting of the holographic plate is very sensitive and delicate. In the digital technique, energy at 1 µJ is sufficient to illuminate the sensor (155 lines per millimeter in resolution). The implementation is easy enough and the phase difference is entirely estimated with a computer. The coherence lengths of the three lasers must be more than 2 m in two optical setups. In silver-halide holographic interferometry, approximately 220 successive frames of the phenomenon can be recorded at high frame rate (35,000 images per second with an exposure time of 750 ns for each). Each image has to be digitalized and processed. Also, it is important to obtain a reference hologram with approximately 50% diffraction efficiency for the three lines. In digital holographic interferometry, the frame rate is limited to 9 frames per second, full size and a synchronized triggering of interferograms recording has to be used to analyze unsteady phenomena.

3.3. Analysis of a supersonic jet with high-density gradients

The feasibility of three-wavelength digital holographic interferometry has been shown for analyzing unsteady wake flows in subsonic flows. Results were obtained from an optical system based on a Michelson interferometer. However, when the refractive index gradients are very strong (e.g. flows encountered in supersonic regime), the light deviation of optical rays induces a shadow effect that is superimposed to the signal of interference microfringes. Phase steps appear and cause a significant loss of information in the interferograms processing. To reduce the shadow effect, a method based on digital three-wavelength holographic interferometry with a single crossing of the test section and different from the Mach–Zehnder configuration is proposed. The optical setup uses two Wollaston prisms where the interfering beams are highly separated and the two prisms are located at the focusing points in a Z optical setup [DES 13]. The optical arrangement makes it easy to introduce three high spatial carrier frequencies that are useful to perform the phase measurement with and without the flow. Fourier spectra obtained on each channel can be filtered easily because the interferograms are recorded with a 3CCD camera. Computed phase difference maps provide refractive index maps with a very good quality and the radial gas density profiles can be extracted assuming the flow to be axisymmetric. A small jet of expanded air is taken as the test case.

3.3.1. *Definition of an optical setup*

The principle of *Z* optical setup using two Wollaston prisms is given in Figure 3.9. The light source is constituted with three different DPSS lasers (red, green and blue at 660, 532 and 457 nm) and uses two spherical mirrors: 250 mm in diameter and 2.4 m in radius of curvature. This arrangement has the particularity of introducing astigmatism because the optical pieces are not quite on the optical axis of the spherical mirrors. The first biprism is located at the focal length of the first spherical mirror so that the two optical rays that return to the second spherical mirror are constituted by parallel light beams. The second spherical mirror refocuses the light beam into the second Wollaston biprism that is mounted “tumble” with the first one.

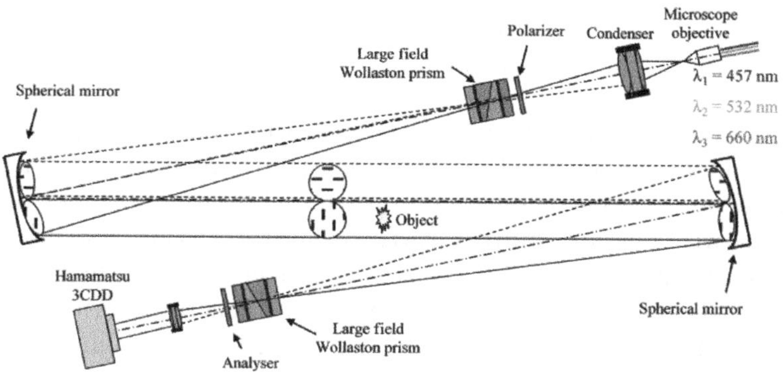

Figure 3.9. *Digital three-color holographic interferometer with two Wollaston prisms. For a color version of this figure, see www.iste.co.uk/picart/digiholography.zip*

An analyzer is then placed behind the second biprism in order to visualize the color interference fringes. A field lens placed in front of the camera forms the image of the object under analysis on the 3CCD sensor. In order to achieve an absolute measurement of the refractive index, we have built specific Wollaston prisms that generate a distance between the two interfering beams greater than the size of the measured field. The interference measurement will be carried out between one reference beam that does not pass through the phenomenon and one other beam that crosses the phenomenon under analysis. Thus, for a spherical mirror, 250 mm in diameter and 2.5 m in radius of curvature (R), the distance dx between the two interfering beams must be ≈125 mm. If $\Delta n = n_e - n_o$ is the birefringence of the crystal and α the bonding angle of prisms, the birefringence angle ε and dx can be expressed according to the following relationships:

$$\begin{aligned}\varepsilon &= \varepsilon(\lambda) = 2(n_e - n_o)\tan(\alpha)\\ dx &= \varepsilon R = 2R(n_e - n_o)\tan(\alpha)\end{aligned} \quad [3.2]$$

By choosing a calcite prism, the bonding angle can be found from the following relation:

$$\alpha = \tan^{-1}\left(\frac{dx}{2R(n_e - n_o)}\right) = \tan^{-1}\left(\frac{0.125}{5\times 0.172}\right) = 8.27° \quad [3.3]$$

From these considerations, large Wollaston prisms were manufactured with an 8° prism angle.

The two beams emerging from the first Wollaston prism are orthogonally polarized. The Z optical path generates astigmatism because all the optical components are not aligned on the optical axis of the two spherical mirrors (off-axis mirrors). The first prism is located at the focal point of the first spherical mirror so that the two optical rays that are reflected onto the second spherical mirror may be parallel. The second spherical mirror refocuses the light into the second Wollaston prism, which is mounted "tumble" with the first one. An analyzer is then placed behind the second prism in order to produce the digital hologram. A field lens located in front of the camera creates the image of the test section at/or nearby the recording plane. The sensor is based on a 3CCD-cooled digital color camera equipped with three CCD chips, 1,344 × 1,024 pixels sized 6.45 µm × 6.45 µm. The colors are separated using a dichroic prism.

Note that due to astigmatism in the setup, the focusing point in front of the sensor is not unique. Figure 3.10 illustrates how the optical beams are focused on the two small focal lines separated by a few millimeters along the optical axis. The first focal line is due to the focusing beam in the horizontal plane (tangential image (TI)), and the second one is the sagittal image (SI) and it is due to focusing beams in the vertical plane.

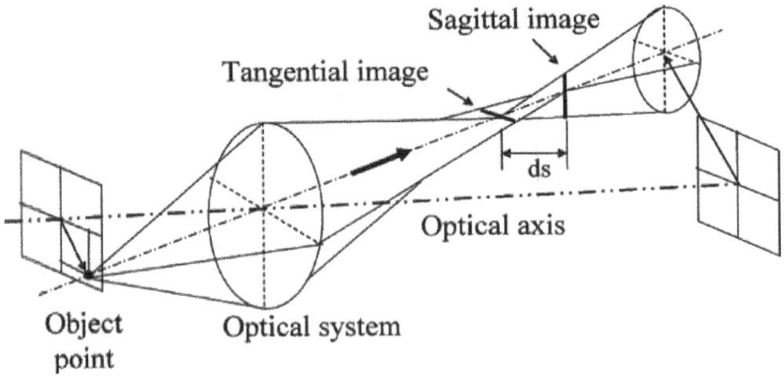

Figure 3.10. *Astigmatism represented by sectional views*

The interference fringes are shown in Figure 3.11.

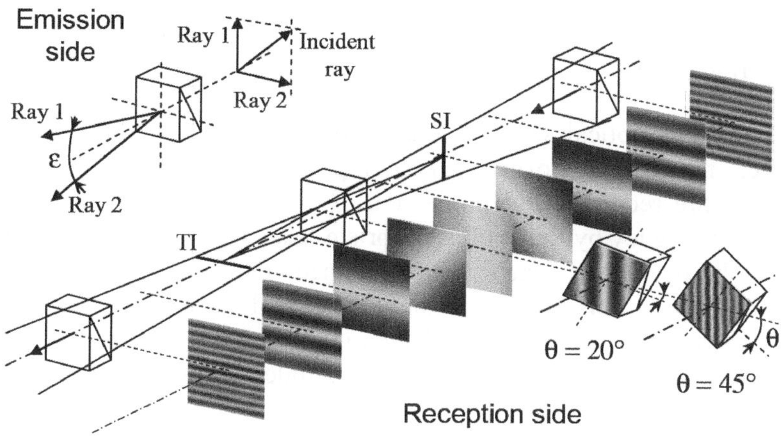

Figure 3.11. *Evolution of interference fringes when the second Wollaston is moved from the sagittal image to the tangential image. For a color version of this figure, see www.iste.co.uk/picart/digiholography.zip*

For example, on the emission side, the three incident rays are polarized at 45° of the crystal axis and the first Wollaston prism is located so that the two interfering rays are split in a vertical plane. On the reception side, when the second Wollaston prism is successively moved along the optical axis toward the SI, the interference fringes, which were horizontal and much narrowed, now spread. When the Wollaston prism is moved from the SI to the TI, the interference fringes spread again, but they rotate by 90° to give a quasi-uniform vertical background color, at half distance between the tangential and SIs. Then, they continue to rotate by 90° up to the SI and narrow to become horizontal. Above the SI, interference fringes stay horizontal and narrow more. In our experiments, the Wollaston prism is located at half distance between the tangential and SIs, because it allows generating interference fringes in the same direction as the direction of the two interfering beams (vertical shift and vertical fringes). This feature has been widely described by Gontier [GON 66]. In fact, to increase the number of fringes in the visualized field, the Wollaston prism has to be turned on itself in the plane perpendicular to the optical axis. Two positions are shown in Figure 3.11 (20° and 45°) where we can see that the maximum number of fringes is obtained for a rotation of 45°.

From this optical setup, digital color holograms including spatial carrier frequencies (off-axis holography, see Chapter 1, section 1.1.3.1) are recorded and processed. The introduction of the spatial carrier frequency by the Wollaston prism leads to the recording of one color hologram at each instant, since there is no need for phase shifting. This provides a single shot and real-time capability to the experimental setup to investigate high-speed transient flows. After Fourier transforming the result, the amplitude and phase maps can be computed.

3.3.2. *Results obtained with a supersonic jet*

The optical setup was tested to analyze the supersonic flow of a small vertical jet, 5.56 mm in diameter at different pressures of injection. Figure 3.12 shows the digital color holograms for the reference and the measurement with a zoom near the injection area, for a pressure of 4 bars.

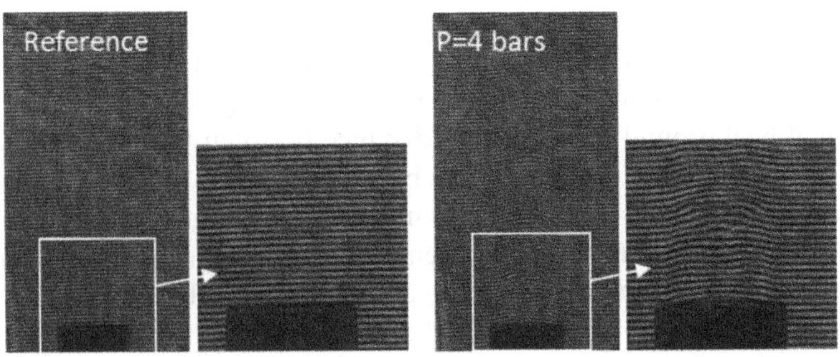

Figure 3.12. *Digital color holograms for a pressure P = 4 bars and zoom in the area near the injection nozzle. For a color version of this figure, see www.iste.co.uk/picart/digiholography.zip*

The spatial carrier fringes, having a horizontal direction, can be clearly seen. In the zoomed images, we can see the horizontal microfringes distorted by the flow. Figure 3.12 also shows that the field of view is reduced on the right and left sides. This is due to the rotation of the Wollaston prism that has a limited square aperture (15 mm²). The two orthogonally polarized beams, which were completely separated on the optical paths, interfere when the prism and the analyzer placed in front of the camera are rotated around its optical axis. Figure 3.13 shows the three R–G–B Fourier spectra calculated from the recorded holograms. Since the microfringes are horizontally oriented, the +1 order is localized along the vertical axis. The filtering is represented by a colored circle and the diameters of the filters are, respectively, $\Delta u = \Delta v = 0.172$ mm^{-1} for the blue channel, $\Delta u = \Delta v = 0.137$ mm^{-1} for the green and $\Delta u = \Delta v = 0.126$ mm^{-1} for the red. Thus, the spatial resolutions in the field of view are, respectively, 25.48, 21.31 and 19.6 lines/mm.

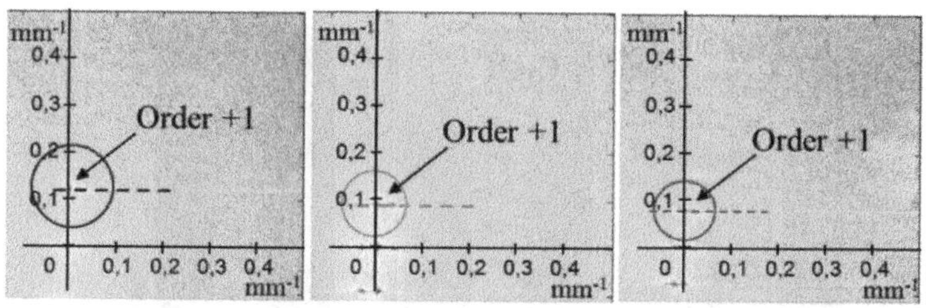

Figure 3.13. *2D Fourier spectra calculated on blue, green and red channels at P = 4 bars. For a color version of this figure, see www.iste.co.uk/picart/digiholography.zip*

Figure 3.14 shows the phase map for the blue channel, for the reference and for a pressure of 3 bars. The spatial carrier frequency remains in data but will be removed when computing the phase difference between the phase at a given pressure and the reference one.

Figure 3.15 shows modulo 2p (a) and unwrapped (b) phase differences obtained at the same pressure for the green channel. The unwrapped phase scale covers approximately 15 radians. The optical thickness is calculated from the optical variations as

$e = \lambda\Delta\varphi_\lambda/2\pi$. Figure 3.15(c) indicates that the variation between the structure's core and the air outside of the jet is 0.6 μm.

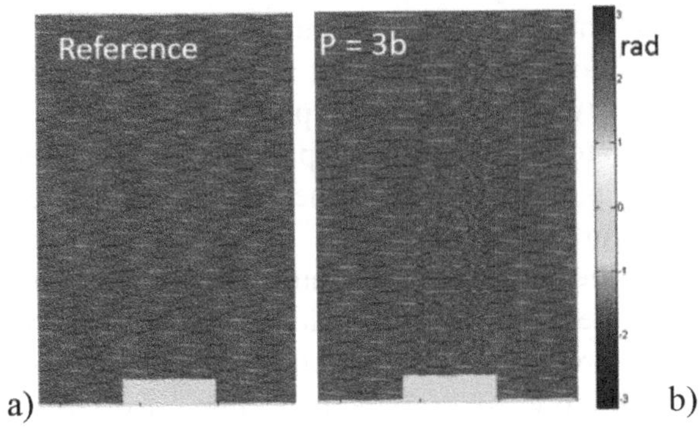

Figure 3.14. *a) Modulo 2π reference phase and b) modulo 2π phase for P = 3 bars. For a color version of this figure, see www.iste.co.uk/picart/digiholography.zip*

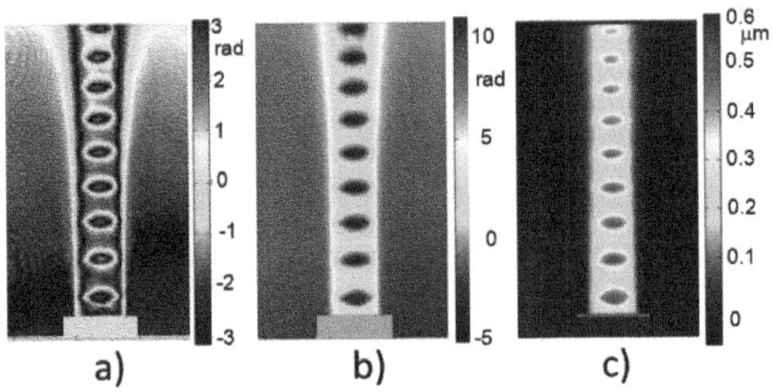

Figure 3.15. *Modulo 2π phase difference a), unwrap phase difference b) and optical thickness c), for P = 3 bars. For a color version of this figure, see www.iste.co.uk/picart/digiholography.zip*

Figure 3.16(a) shows the superimposition of the three interferograms obtained from the phase variations for a pressure at 5 bars. The color interference fringes are those which should be obtained if color holographic interferometry using holographic plates would be implemented. From the optical thickness map, the radial distribution of the refractive index, assuming that the flow is axisymmetric (Figure 3.16(b)), can also be extracted.

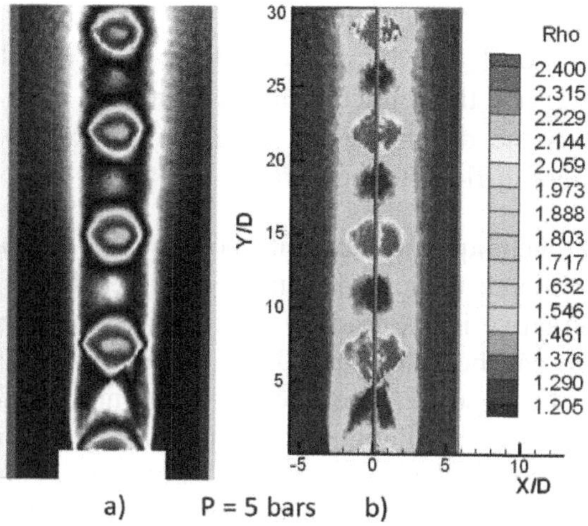

Figure 3.16. *a) Color luminous intensity and b) radial gas density at P = 5 bars. For a color version of this figure, see www.iste.co.uk/picart/digiholography.zip*

The analysis is performed from the left and right parts of the jet and it is stopped on the jet axis [ROD 97]. The gas density profiles can be found in the jet, according to the Gladstone–Dale relationship.

In conclusion, an optical setup based on three-wavelength digital holographic interferometry using Wollaston prisms is presented and includes several advantages compared to other configurations: single crossing of the test section, lower sensitivity to the shadow effect, adjustable spatial carrier frequencies both in amplitude and orientation and Fourier data processing, which means real-time capability for the setup. Experimental proof of concept has been performed on a microjet. Assuming the flow is axisymmetric, gas density profiles can be extracted from the optical thickness maps.

3.4. Analysis of a hydrogen jet in a hypersonic flow

The implementation of digital holographic interferometry has been considered in several ONERA wind tunnels among which the F4 hypersonic Fauga–Mauzac facility. Here, the difficulty lies in the fact that as the gas density of the upstream flow ρ_0 is very low (close to 10^{-4} or 10^{-5} kg/m^3), the refractive index n_0 related to ρ_0 is also very small, and thus the differences in refractive index $(n-n_0)$ are even smaller and more difficult to measure.

The optical technique used up until now in the F4 hypersonic wind tunnel is a schlieren technique using a schlieren knife or a colored mask. The refractive index gradients in the direction perpendicular to the knife edge can be visualized. Thus, if the knife edge is located vertically in the focusing spot beam, horizontal gradients in the refractive index will be highlighted. This technique shows its limits when the refractive index gradients are very low. In this case, it is better to use optical methods that directly visualize the refractive index and not the refractive index gradient. Holography and holographic interferometry are suitable to measure the absolute value of the refractive index and holographic interferometry has already been exploited in the F4 hypersonic facility, in which the crossing number of the test section was multiplied by 2, 3 or 4, which increased in the same ratio of the measurement sensitivity [SUR 92a, SUR 92b]. A comparison between the schlieren technique and transmission holographic interferometry using a holographic plate was shown. The difficulty with the transmission holographic interferometry developed by Surget was that the noise increased with the crossing number, which makes the quantitative processing of interferograms difficult. Otherwise, the idea, which consists of crossing the flow several times to increase the measurement sensitivity, is an idea that can be applied to digital holographic interferometry, based on the principle described by Surget and Dunet.

3.4.1. *Experimental setup*

The ONERA F4 wind tunnel, commissioned during the course of the ESA HERMES program in the early 1990s, has since been extensively used to test various re-entry vehicles, for both terrestrial and Martian atmosphere [VIG 12]. It is a hotshot-type facility, meaning that the operating conditions are obtained by heating the test gas with an intense electric arc in a chamber of 10–15 L volume initially pressurized at ambient temperature. This chamber is a cylindrical tank whose lateral sides are the electrodes; the electrode shape induces a rotation of the arc. This enhances the efficiency of the heating process and reduces the erosion of the electrode head. The electrode bodies are thermally and electrically insulated from the chamber. Two exits connect the chamber downstream of the nozzle and upstream of the quick discharge exhaust tank. The energy is delivered by an impulse generator, at a power of up to 150 MW for several tens of milliseconds. The settling-chamber pressure Pi can be as high as 1 kbar, and the total enthalpy can be as high as 18 MJ/kg for air. Run duration of up to 400 ms can be achieved.

The aim of this test is to determine if the digital holographic interferometry can be used to visualize the injection of a small jet of hydrogen at 32 bar on a flat plate at a 20° incidence angle. If the jet is correctly visualized with the optical technique, information on the combustion process of the hydrogen jet in the shock layer can be obtained. In this test, the plate is 260 mm wide and the injection is located at 117 mm from the leading edge, as shown in Figure 3.17.

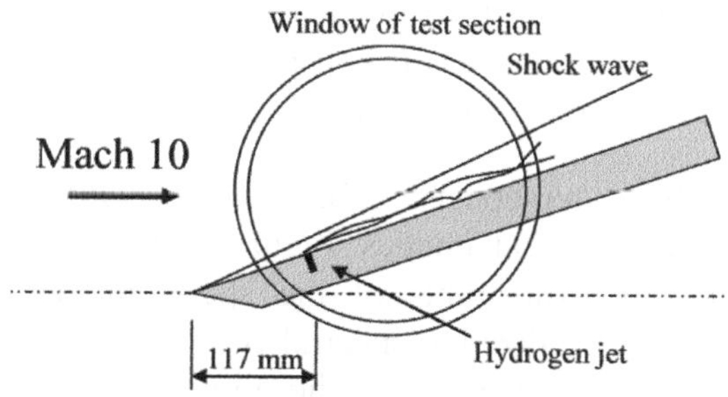

Figure 3.17. *Schematic model*

The optical setup looks like that of Figure 3.3 but just one DPSS laser ($\lambda = 532$ nm) is used as the light source of the interferometer.

3.4.2. *Experimental results*

As the test duration is approximately 200 ms, the interferograms are recorded with a Photron APX RS 3000 camera at a frame rate of 3,000 images per second (333.33 µs between each interferogram). The exposure time of each image fixed by the camera is 1 µs and the camera is triggered by a Transitor-Transistor Logic (TTL) signal generated 215 ms before the beginning of the burst. The recording procedure is as follows:

– the interferometer is adjusted with a uniform background color. Then, the concave mirror is tilted to introduce the reference microfringes;

– the reference interferogram is recorded without flow in the test section;

– finally, the camera triggered with the run records the measurement interferograms at a high framing rate. Here, as the camera is triggered a few hundred microseconds before the beginning of the burst, the first images of the film can easily be used as the reference images.

Figure 3.18(a) shows two images of interference microfringes recorded with and without flow (reference and measurement). In the enlarged part of the shock layer located just above the injection, the disturbance of the interference microfringes can be observed, both by the shock wave and the hydrogen jet.

The first analysis step consists of computing 2DFFT spectra of the reference and measurement images. Figure 3.18(b) shows the results obtained in the four quadrants of the spectrum. The +1 and –1 symmetrical orders (localized at spectral coordinates corresponding to the carrier frequencies generated by the optical setup) and the central zero order are very well identified. The filtering step consists of applying a window around the +1 order and in reconstructing the amplitude and phase of the hologram with the information contained around the +1 order.

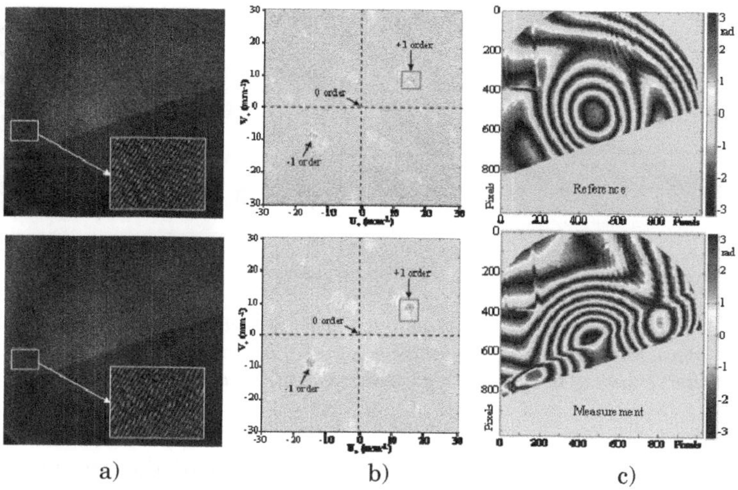

Figure 3.18. *Reference and measurement interferogram analysis: a) interferograms, b) Fourier spectra and c) modulo 2π phase difference. For a color version of this figure, see www.iste.co.uk/picart/digiholography.zip*

Considering the values of the spatial carrier frequencies, u_0 and v_0 are, respectively, equal to 15 and 10 mm^{-1} and give a resolution of approximately 10–15 fringes per millimeter in the sensor plane. As the optical setup magnification is here given by the ratio between the diameter of the large achromatic lens (200 mm) and the sensor size (17 mm), the resolution in the test section is equal to 0.76 mm along the x-axis and 1.14 mm along the y-axis. These values may seem high, but they are mainly related to the pixel size of the camera that was used in this feasibility study. Once the filtering window is determined, the reference and measurement phase maps shown in Figure 3.18(c) are calculated. It should be noted that the phase maps can also be calculated by saving the spatial carrier frequency.

As the phases are calculated modulo 2π, an unwrapping process has to be applied to obtain the phase map. Figure 3.19(a) shows the phase processing (unwrapped) for the measurement. From the phase map of the phenomenon, the optical thickness map can easily be calculated and Figure 3.19(b) shows an example where the optical thickness varies between –0.4 and 0.4 μm.

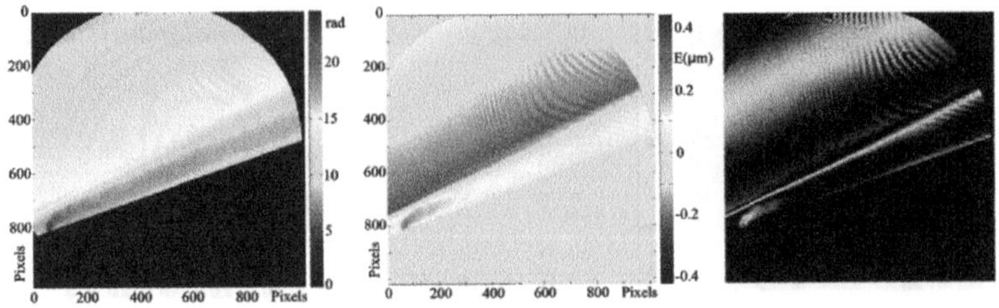

a) Unwrapped phase difference b) Optical thickness c) Luminous intensity

Figure 3.19. *Visualization of the hydrogen jet: a) unwrapped phase difference, b) optical thickness and c) luminous intensity. For a color version of this figure, see www.iste.co.uk/picart/digiholography.zip*

After having analyzed all the interferograms of the sequence, we note that no evidence of combustion of the hydrogen jet in the shock layer can be pointed out. Knowing the phase difference of each frame, the luminous intensity of the interference fringes, which would have been visualized if a technique using transmission or reflection plate holography had been implemented, can be calculated. Figure 3.19(c) shows the interferogram computed from the phase difference maps. The subtraction of an initial optical thickness is equivalent to recording interferogram with a uniform background color, which leads to a fine visualization of the shock wave location and the small hydrogen jet injection.

3.4.3. *Comparisons with numerical simulations*

The wind-tunnel run studied in previous sections can also be numerically simulated at an instant of the run slightly earlier than for holographic visualization so that the reservoir total pressure (Pt) and the enthalpy (Ht) weakly differ: Pt = 497 bar and Ht = 9.34 MJ/kg for the computation; Pt = 472 bar and Ht = 8.70 MJ/kg for the visualization. Nevertheless, the measured Pitot pressures (P't) in the test section are very close: P't = 27,058 Pa for the computation and P't = 27,844 Pa for the visualization.

3D Reynolds-averaged Navier–Stokes (RANS) computations have been performed to simulate the flow. Only the flow on the upper part of the plate is computed. The geometry of the injector is not modeled: injection is simulated by a step velocity profile imposed on a disk boundary condition of 1 mm diameter. The mesh used is block-structured and contains approximately 10.4 million cells. The cell size at the injection area is 50 µm and the height of the first cell above the wall is 20 µm.

Starting from the reservoir pressure and enthalpy, the flow is rebuilt inside the test section using the PNS code PANASCE. At $t = 35$ ms, the Mach number, static pressure and static temperature in the test section are, respectively: 9.58, 246 Pa and 471 K. These values are used as input for the RANS computation. The computed unit Reynolds number inside the test section is 274,262 m^{-1}. Hydrogen is injected at Mach 1 with a total pressure of 17.4 bar and a total temperature of 300 K. Computations are run with the ONERA CEDRE code. Turbulence is taken into account using a *k*–l model. Combustion is modeled by a quasi-laminar approach: terms of species production are calculated by the Arrhenius law based on the Eklund kinetics [EKL 90]. Eulerian fluxes are evaluated at cells faces using a Roe scheme. Time integration is performed by an implicit Euler scheme of first-order accuracy. Second order in space is achieved by a Monotonic Upstream-Centered Scheme for Conservation Laws (MUSCL)-type interpolation on cells faces.

Principal features of the flow are gathered in Figure 3.20, which focus on the injection [SCH 66, VIT 09]. The jet expands through an expansion fan and is recompressed through a barrel shock whose windward part is folded by the transverse main flow. The bow shock created in front of the injection intersects the plate leading edge shock. The adverse pressure gradient induced by this bow shock causes boundary layer separation, which results in a characteristic λ shape shock pattern. The barrel shock ends with a Mach disk after which flow is subsonic. This subsonic flow is then reaccelerated by surrounding flow through mixing layers. Another mixing layer takes place between the bow shock and the barrel shock. Even if the detail of the flow is not captured by holographic visualization, the interaction between the leading-edge shock and the injection bow shock is clearly visible in Figure 3.20(a). In Figure 3.20(b), the schlieren visualization obtained from holographic interferogram is compared to a "numerical" schlieren computed from simulation, in the plane of symmetry. Due to the fact that the numerical schlieren is calculated in the symmetry plane, the leading-edge shock cannot be captured downstream of its interaction with the jet bow shock. For the same reason, the bow shock itself is more discernible than in the holographic-based visualization. Nevertheless, the comparison between the two approaches shows a very good agreement on the location of the shocks interaction, and of the downstream position of the bow shock.

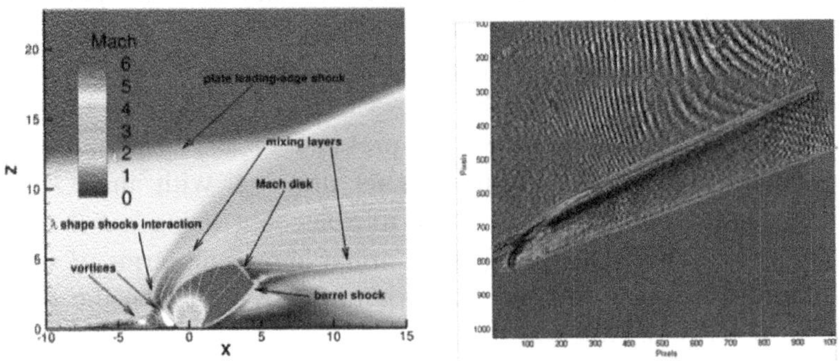

a) b)

Figure 3.20. *2D view in the symmetry plane of the flow structure near the injection a), and comparison of schlieren visualizations from holographic measurement and from numerical simulation b). For a color version of this figure, see www.iste.co.uk/picart/digiholography.zip*

3.5. Conclusion

As a conclusion, some improvements can be brought to this optical setup. As the resolution of digital holographic interferometry depends mainly on the pixel size and the pixel number of the CCD or Complementary Metal Oxide Semiconductor (CMOS) sensor, the interferograms have to be recorded with a camera having a high spatial resolution sensor and a high frame rate. In order to increase the spatial carrier frequency, the wavelength of the green DPSS laser can be replaced by a blue DPSS laser having an adapted coherence length. The sensitivity of the measurement can be increased by changing the focal length of the large achromatic lens (800 mm) and the optical bench has to be isolated from external vibrations.

As such, the feasibility of digital holographic interferometry has been demonstrated in the F4 hypersonic wind tunnel. The phase difference maps allow the optical path difference, or thickness maps, to be obtained if the flow is assumed to be 2D in the span wise direction of the flat plate. The refractive index maps of the phenomenon could be obtained from the optical thickness maps.

The results obtained with digital holographic interferometry can be used for several interesting comparisons. From the phase difference maps, the holographic images and schlieren images in the x and y directions can be recalculated, which is not possible with a conventional schlieren optical setup.

Finally, the feasibility of the method has shown the broad possibilities of the technique and gives access to the quantitative measurements of the flow as the optical thickness and the refractive index under particular conditions. In addition, if the characteristics of the ionized gas are well known, the value of the gas density upstream and in the shock layer can be estimated.

In the near future, digital three-wavelength holographic interferometry seems to be the best candidate to characterize unsteady complex flows. Although CCD resolution and size are not as good as that of holographic plates, the digital approach is more accessible and versatile since the time for the hologram processing is greatly reduced and the processing is purely numerical. On the other hand, the value of using color has been demonstrated as the zero-order fringe can be easily determined and the variation in the background color due to disturbances can be quantified.

3.6. Acknowledgment

The author thanks the French National Agency for Research (ANR) since part of this research was partially funded by grant no. ANR 2010 BLAN 0302.

3.7. Bibliography

[DEM 03] DEMOLI N., VUKICEVIC D., TORZYNSKI M., "Dynamic digital holographic interferometry with three wavelengths", *Optics Express*, vol. 11, pp. 767–774, 2003.

[DES 97] DESSE J.M., "Three-color differential interferometry", *Applied Optics*, vol. 36, no. 28, pp. 7150–7156, 1997.

[DES 03] DESSE J.M., "Oil-film interferometry skin-friction measurement under white light", *AIAA Journal*, vol. 41, pp. 2468–2677, 2003.

[DES 08] DESSE J.M., PICART P., TANKAM P., "Digital three-color holographic interferometry for flow analysis", *Optics Express*, vol. 16, no. 8, pp. 5471–5480, 2008.

[DES 09] DESSE J.M., TRIBILLON J.L., "Real-time three-color holographic interferometry", *Applied Optics*, vol. 48, pp. 6870–6877, 2009.

[DES 12] DESSE J.M., PICART P., TANKAM P., "Digital color holography applied to fluid and structural mechanics", *Optics and Lasers in Engineering*, vol. 50, pp. 18–28, 2012.

[DES 13] DESSE J.M., PICART P., "Digital three-wavelength holographic interferometry using Wollaston prisms", *Proceedings of the Digital Holography and Three-Dimensional Imaging, OSA Technical Digest*, Optical Society of America, paper DTh4A.1, 2013.

[EKL 90] EKLUND D.R., DRUMMOND J.P., HASSAN H.A., "Calculation of supersonic turbulent reacting coaxial jets", *AIAA Journal*, vol. 28, no. 9, pp. 1633–1641, 1990.

[FER 04] FERRARO P., DE NICOLA S., COPPOLA G., *et al.*, "Controlling image size as a function of distance and wavelength in Fresnel-transform reconstruction of digital holograms", *Optics Letters*, vol. 29, pp. 884–886, 2004.

[FER 07] FERRARO P., MICCIO L., GRILLI S., *et al.*, "Quantitative phase microscopy of microstructures with extended measurement range and correction of chromatic aberrations by multi wavelength digital holography", *Optics Express*, vol. 15, pp. 14591–14600, 2007.

[GON 66] GONTIER G., CARR P., HENON G., "Sur un dispositif d'interférométrie différentielle permettant d'orienter arbitrairement les franges par rapport à la direction du décalage des faisceaux", *Compte-rendu Académie Sciences*, vol. 262, pp. 674–677, 1966.

[JAV 05] JAVIDI B., FERRARO P., HONG S., *et al.*, "Three-dimensional image fusion by use of multiwavelength digital holography", *Optics Letters*, vol. 30, pp. 144–146, 2005.

[JEO 97] JEONG T.H., BJELKHAGEN H.I., SPOTO L.M., "Holographic interferometry with multiple wavelengths", *Applied Optics*, vol. 36, pp. 3686–3688, 1997.

[KAT 02] KATO J., YAMAGUCHI I., MATSUMURA T., "Multicolor digital holography with an achromatic phase shifter", *Optics Letters*, vol. 27, pp. 1403–1405, 2002.

[KUN 07] KUHN J., COLOMB T., MONTFORT F., *et al.*, "Real-time dual-wavelength digital holographic microscopy with a single hologram acquisition", *Optics Express*, vol. 15, pp. 7231–7242, 2007.

[PIC 03] PICART P., MOISSON E., MOUNIER D., "Twin sensitivity measurement by spatial multiplexing of digitally recorded holograms", *Applied Optics*, vol. 42, pp. 1947–1957, 2003.

[PIC 05] PICART P., LEVAL J., GRILL M., *et al.*, "2D full field vibration analysis with multiplexed digital holograms", *Optics Express*, vol. 13, pp. 8882–8892, 2005.

[ROD 97] RODRIGUEZ O., DESSE J.M., PRUVOST J., "Interaction between a supersonic hot jet and a coaxial supersonic flow", *Aerospace Science and Technology*, vol. 11, no. 6, pp. 369–379, 1997.

[SCH 66] SCHETZ J.A., HAWKINS P.F., LEHMAN H., "Structure of highly underexpanded transverse jets in a supersonic stream", *AIAA Journal*, vol. 5, no. 5, pp. 882–884, 1966.

[SCH 94] SCHNARS U., JÜPTNER W., "Direct recording of holograms by a CCD target and numerical reconstruction", *Applied Optics*, vol. 33, pp.179–181, 1994.

[SUR 92a] SURGET J., PHILBERT M., BIZE D., "Schlieren device and holographic interferometer for hypersonic flow visualization", *Proceedings of the 6th International Symposium on flow Visualization*, Yokohama, Japan, 1992.

[SUR 92b] SURGET J., DUNET G., "Multipass holographic interferometer for the high enthalpy hypersonic wind tunnel F4", *Proceedings of the Advanced Research Workshop on New Trends in Instrumentation for Hypersonic Research*, Le Fauga, France, 1992.

[VIG 12] VIGUIER P., GARRAUD, J., SOUTADE, J., *et al.*, "Development of F4 hotshot windtunnel for high enthalpy scramjet tests", *Proceedings of the 18th AIAA/3AF International Space Planes and Hypersonic Systems and Technologies Conference*, Tours, France, 2012.

[VIT 09] VITI V., NEEL R., SCHETZ J.A., "Detailed flow physics of the supersonic jet interaction flow field", *Physics of Fluids*, vol. 21, no. 4, pp. 1–16, 2009.

[YAM 97] YAMAGUCHI I., ZHANG T., "Phase-shifting digital holography", *Optics Letters*, vol. 22, pp. 1268–1270, 1997.

[YAM 02] YAMAGUCHI I., MATSUMURA T., KATO J., "Phase-shifting color digital holography", *Optics Letters*, vol. 27, pp. 1108–1110, 2002.

[ZHA 08] ZHAO J., JIANG H., DI J., "Recording and reconstruction of a color holographic image by using digital lensless Fourier transform holography", *Optics Express*, vol. 16, pp. 2514–2519, 2008.

4

Automation of Digital Holographic Detection Procedures for Life Sciences Applications

4.1. Introduction

Ever since the first use of the microscope in the early 17th Century, the field of optical microscopy has been widely investigated and the performances continuously improved. Over the last 20 years, the development of new technologies breathed new life into the field of optical microscopy by pushing the boundaries of the capabilities. Nevertheless, classical optical microscopy is still limited by the small depth of focus due to the high magnification and the high numerical apertures (NAs). The extension of the depth of focus has become an important goal over the last decade. With the recent developments in digital cameras, digital holographic microscopy (DHM) provided an elegant and natural solution to the problem by providing to ability to reconstruct numerically different slides in an experimental volume [SCH 02]. In this way, the depth of focus has been drastically increased. Furthermore, due to the interferometric nature of the technique and depending on the optical setup, DHM provides a quantitative access to the optical phase which has widened considerably the field of investigation, especially in biological and

Chapter written by Ahmed EL MALLAHI, Christophe MINETTI and Frank DUBOIS.

environmental applications [MIN 08, ELM 13, DOH 14, YOU 14]. However, with the usual laser sources, the increase of the depth of focus has important consequences on the image quality because of the speckle noise inherent to coherent sources. Although, the impact of this noise is reduced when working only with the optical phase (like in quantitative phase contrast imaging), it becomes important and non-negligible when using the complex amplitude (intensity and phase) in the numerical implementation of the optical field propagation. For the last decade [DUB 99], we have developed optical setups working with spatially reduced coherent sources. The use of reduced coherence strongly improves the image quality and consequently the in-depth numerical reconstruction [DUB 04].

Digital holography with reduced coherent sources can be implemented in different ways. Among those, the off-axis configuration is particularly suitable for dynamic phenomenon as the acquisition of a single frame provides the full complex amplitude [KRE 86]. It allows us to work with flux cells collecting a large amount of information on the sample in a short span of time, which is convenient for biological or environmental applications where collective effects play an important role [DUB 06a, DCA 14, YOU 14]. For example, 1 μl of human blood contains around five million red blood cells (RBCs). This large amount of particles requires, to be realistic, a fully automated processing of the holograms to extract data of interest. This implies, at the very first step, the need for detection methods to locate, in the experimental volume, the different objects. In most of the cases, this detection is not straightforward as it is problem-dependent and the in-depth dimension brings an additional complexity to the problem. In this chapter, we propose automated detection methods exploiting the full interferometric information.

We will first describe the experimental protocol with the optical setup and the micro-fluidic device. Before presenting the detection methods, we will describe specific tools linked to the numerical propagation that are indispensable to process the information correctly, avoid numerical effects and make easier the further processing. The automated three-dimensional (3D) detection methods based on propagation matrices will then be developed with a local and a global approach. Advantages and limitations of the different approaches will be discussed and illustrated using concrete applications.

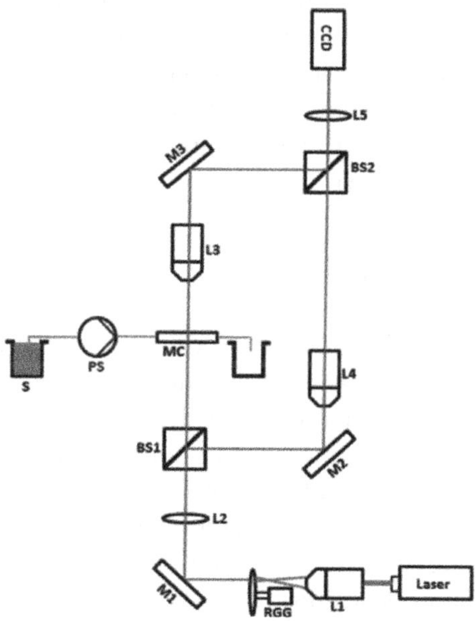

Figure 4.1. *Optical setup of the DHM. For a color version of this figure, see www.iste.co.uk/picart/digiholography.zip*

4.2. Experimental protocol

4.2.1. *Optical setup*

The optical setup used is shown in Figure 4.1. A coherent beam emerging from a mono-mode laser diode ($\lambda = 635$ nm) is focalized by lens L1 on a rotating ground glass (RGG). The RGG decreases the coherence of the source by placing it on the optical axis close to the focus point of lens L1. In this way, we can adjust the spatial size of the source easily and consequently its spatial coherence. The light transmitted by the RGG is scattered and collimated by lens L2. A beam splitter BS1 divides the collimated beam in an object and a reference beam. The object beam passes through the sample (MC) and the microscope lens L3 to provide magnification. In a symmetric way, the reference beam passes through the microscope lens L4 identical to L3. The reference and object beams are then recombined by a second beam splitter (BS2) and transmitted by lens L5. Lenses L3-L5 image one plane of the sample onto the charge-coupled device (CCD) camera sensor. Then, the magnification G_S is given by the ratio between the focal length L5 and the focal length L3. Depending on the applications, the magnification is adjusted by changing the couple of lenses L3 and L4 (typically, we use microscope lenses of ×10, ×20, ×40 and ×63 magnification).

4.2.2. *Dynamic monitoring*

The experimental cell is made of a microchannel connected to an injection system via an inlet and an outlet to create a dynamic flux. The dimensions of the microchannel depend on the application and typically range from 20 μm to several hundreds of μm in a square or rectangular section. The fluid sample is injected or pumped into the microchannel via a push/pull programmable pump. Flow rates and exposure times are adjusted to avoid any blurring effects of the particles. A camera is acquiring holograms at a frequency of 24 frames per second. Due to the spatial coherence reduction of the source, home-made microchannels that are not especially optimized with respect to the optical quality can be used.

4.3. General tools

In this section, we present general tools specific to digital holography that will be widely used for the detection.

4.3.1. *Extraction of the full interferometric information*

Once a digital hologram is recorded, the complex amplitude is first extracted using the Fourier method [KRE 86]. The intensity and the phase are then computed. These two quantities constitute the full interferometric information.

4.3.2. *Compensation of the phase*

The complex nature of the optical field provides quantitative intensity and phase maps. The quantitative phase information is particularly important for purely phase objects like in biological applications. Due to the periodicity of the imaginary exponential, phase maps usually present phase jumps. This non-uniform phase background has to be removed in order to exploit the quantitative aspect of the phase information more easily. Due to the symmetry between the object and the reference beam of the interferometer, we implemented a fast phase map compensation method that consists of fitting the background phase with a bi-dimensional parabolic phase map. In many situations, this phase map compensation is sufficient to correct the phase errors. Let us consider that the acquired phase $\psi(x,y)$ is the sum of the phase induced by the object $\varphi(x,y)$ and the background phase $\beta(x,y)$ that corresponds to the phase obtained when no object is present in the optical system:

$$\psi(x,y) = \phi(x,y) + \beta(x,y) \qquad [4.1]$$

The background phase can be approximated by a quadratic phase map:

$$\beta(x,y) = Ax^2 + By^2 + Cx + Dy + E \qquad [4.2]$$

where A, B, C, D and E are parameters to be determined. The fitting of the background phase cannot be implemented directly due to phase jump. To avoid the time-consuming unwrapping of thc phase, we consider the derivatives of the phase in the x- and y-directions:

$$\begin{aligned} \frac{\partial \Psi(x,y)}{\partial x} &= \frac{\partial \phi(x,y)}{\partial x} + 2Ax + C \approx 2Ax + C \\ \frac{\partial \Psi(x,y)}{\partial y} &= \frac{\partial \phi(x,y)}{\partial y} + 2By + D \approx 2By + D \end{aligned} \qquad [4.3]$$

where we made the assumption that phase changes induced by the objects are small compared to the phase changes induced by the background phase. In this way, the problem of fitting the quadratic phase map is reduced to the fitting of two planes in the x- and y-directions. Let us define those two planes as follows:

$$\begin{aligned} \beta_x(x,y) &= 2Ax + C \\ \beta_y(x,y) &= 2By + D \end{aligned} \quad [4.4]$$

We define the mean square error:

$$E_x = \sum_{x,y=1}^{N} \{\beta_x - (2Ax + C)\}^2 \quad [4.5]$$

Minimizing the mean square error with respect to the parameters A and C gives us:

$$\begin{aligned} \frac{\partial E_x}{\partial A} &= \sum_{x,y=1}^{N} \left(x \cdot \beta_x(x,y) - 2Ax^2 - Cx\right) = 0 \\ \frac{\partial E_x}{\partial C} &= \sum_{x,y=1}^{N} \left(\beta_x(x,y) - 2Ax - C\right) = 0 \end{aligned} \quad [4.6]$$

which can be rewritten as:

$$\begin{cases} 2A \cdot \sum_{x,y=1}^{N} x^2 + C \cdot \sum_{x,y=1}^{N} x = \sum_{x,y=1}^{N} x \cdot \beta_x(x,y) \\ 2A \cdot \sum_{x,y=1}^{N} x + C \cdot \sum_{x,y=1}^{N} 1 = \sum_{x,y=1}^{N} \beta_x(x,y) \end{cases} \quad [4.7]$$

Resolving this system of equations gives the values of A and C. Similarly, we define E_y and determine the values of B and D. The offset E remains undetermined and must be fixed arbitrarily. The value of the offset will give the value of the background of the compensated phase. An illustrative example (phospholipid vesicles in an aqueous solution) of compensation is shown in Figure 4.2 with (a) the original phase map $\psi(x,y)$, (b) the fitted background phase $\beta(x,y)$ and (c) the compensated phase $\varphi(x,y)$.

Figure 4.2. *Example of phase map compensation. a) Original phase map $\psi(x,y)$ of a suspension of phospholipid vesicles in an aqueous solution, b) the fitted background phase $\beta(x,y)$ and c) the compensated phase $\varphi(x,y)$ with an offset value of 90 (in the grey level scale)*

This compensation method is very fast and easy to implement. As long as the phase change induced by the objects remains below 2π, the method efficiently removes the background phase and provides phase images with a flat and almost constant background.

4.3.3. *Border processing*

The basis for numerical hologram reconstruction is given by the well-known Kirchhoff–Fresnel (KF) integral which determines, on the basis of the complex amplitude in one plane, the complex amplitude in a plane parallel at a distance *d*. In digital holography, the Kirchhoff–Fresnel integral is implemented numerically for small angles of diffraction in the Fresnel approximation [GOO 05]. This numerical implementation gives rise to border diffraction that has to be processed adequately to keep a good quality of the in-depth reconstruction. Indeed, the discrete implementation of the KF propagation uses the two-dimensional (2D) discrete Fourier transformations (DFTs) which assumes that the 2D amplitude field is sampled both from finite length and cyclic. Consequently, when the border points on each vertical or horizontal line have unequal amplitudes, they are processed as a step function and the discrete KF propagation gives rise to the typical amplitude oscillation of the near-field diffraction. As for the near-field diffraction, this effect becomes more important with the increased propagation distance and can rapidly disturb a large area of the reconstructed image. To remove this effect, we extend the size of the digital hologram with values that minimize the diffraction [DUB 02]. After propagation, we crop the amplitude and phase to their original size. In Figure 4.3, we show an RBC recorded with a defocus distance of 39 µm (a), the propagated intensity to the focus plane of the RBC without (b) and with (c) border processing.

a) b) c)

Figure 4.3. *Example of the border processing. a) A red blood cell recorded with a defocus distance of 39 µm; b) reconstructed intensity at a distance of –39 µm without border processing and c) with border processing*

The border processing drastically reduces the border diffraction influence on the reconstruction. This processing is indispensable as we will see later for a correct best focus determination.

4.3.4. *Best focus determination*

Digital holography permits the numerical reconstruction of an experimental volume but does not give any criterion when the focus distance of the objects inside this experimental volume is reached. The classical autofocus methods used in photo cameras most of the time fall into default when applied to digital holograms. This is because, in digital holography, each reconstructed plane inside the experimental volume contains the same amount of information as the acquired digital hologram; the digital refocusing does not make any filtering of the data.

Among the different methods proposed in the literature [MAR 05, LAN 08], we have proposed ([DUB 06b]) a focus criterion that takes benefit of the invariance of both energy and amplitude integration.

These invariance properties facilitate building two focus criteria, respectively, for pure amplitude and pure phase objects that are based on the score of the integrated amplitude modulus. It can be demonstrated that the integrated amplitude modulus is at its maximum when the best focus plane of a purely phase object is reached and at its minimum for a purely amplitude object. Examples of best focus curves are given in the next section.

This criterion is based on an integral and gives a value that is influenced by all the objects distributed in the volume. If two objects are present in the region of investigation, the z value will be influenced by both objects and will give a mean value which does not correspond to the z position of any of the two objects (unless they are in the same plane). For this reason, the in-depth position must be determined carefully by selecting a zone of integration containing only one object. When the size of the objects is known and fixed (like RBCs for example), the in-depth detection can be performed locally within a region of interest (ROI) cropped around each cell, the integral being computed within the ROI. When it is not the case, the problem becomes more difficult and pre-segmentation of the objects is necessary in order to determine the size of the ROI to be used. As the size of the object might be influenced by the diffraction rings when this one is out of focus, special detection techniques, such as the propagation matrices, need to be implemented; this will be described in the next section. In any case, the border processing is mandatory as it might strongly influence the z value if border diffraction appears in the integration zone.

4.4. Automated 3D detection

4.4.1. *Introduction*

The objective of this chapter is to describe the development of automated procedures that are able to detect and refocus all the particles passing through the field-of-view of a flow-through system in a dynamic acquisition. The purpose is to obtain fully generic methods that could be applied in different kinds of applications; this will be presented in section 4.5. The 3D position determination of objects in DHM is crucial as it constitutes the first step before any further processing.

To achieve the 3D detection, our procedures determine, for the first time, the in-plane coordinates of the objects, i.e. the *x*-*y* positions. Then, the in-depth locations, i.e. the *z* coordinate, are searched. The coordinates *x* and *y* are, respectively, the horizontal and the vertical axis, while the *z*-axis is the optical axis. The 3D detection procedure is thereby divided into two steps:

– in-plane detection (*x*-*y*);

– in-depth detection (*z*).

4.4.2. *Description of the testing samples*

The developed 3D detection procedure is intended to be as general as possible. For this reason, a large variety of species have been tested with the different methods to quantify the performances. Preliminary tests have been performed first on opaque polyethylene spherical particles of 50 μm diameter, as illustrated in Figure 4.4. These synthetic particles are used here to test the different steps of the developed detection procedure before working with actual organisms, which are more complex.

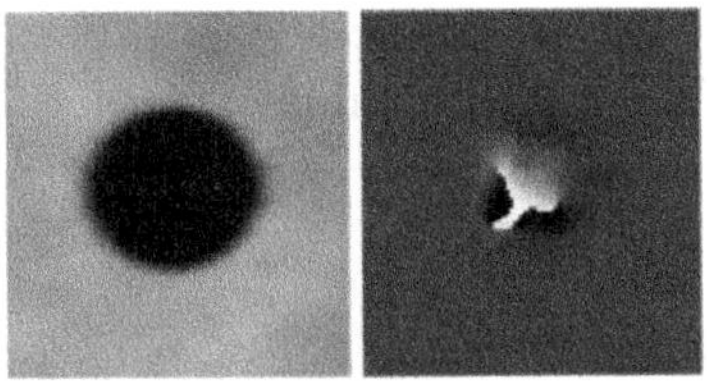

a) b)

Figure 4.4. *An opaque polyethylene spherical particle: a) intensity image and b) compensated phase image*

Indeed, as seen in Figure 4.4(a), there is a high contrast between the particle and the background, which facilitates the detection. Then, experiments with real organisms have been performed. The different organisms tested are illustrated in Figure 4.5.

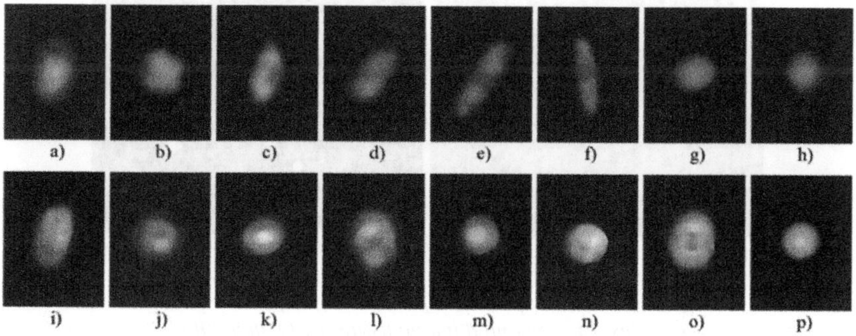

Figure 4.5. *Species used to test the 3D automated detection process – compensated phase images. a) Giardia lamblia, b) Chlorella autotrophica, c) Scenedesmus dimorphus, d) Nitzschia ovalis, e) Nitzschia aurariae, f) Navicula perminuta, g) Cyanothece sp., h) Stanieria sp., i) Rhodomonas saculata, j) Pavlova lutheri, k) Isochrysis galbana, l) Dunaliella tertiolecta, m) Chlorella pyrenoidosa, n) Platymonas convolutes, o) Tetraselmis striata, p) Prasiola sp.*

4.4.3. *In-plane detection*

In this section, we propose two different procedures to achieve in-plane detection of objects, i.e. the determination of the *x*-*y* positions of objects. The first method, called the classical thresholding method, is based on 2D information, while the second method is a 3D method performed through the computation of propagating matrices. These two methods are described in [ELM 13]. In this section, we carry out detailed performance estimations of the two methods.

4.4.3.1. *Classical thresholding method*

4.4.3.1.1. Description of the method

The first category of detection method is based on a classical threshold on the interferometric information extracted by digital holography, i.e. the intensity and the phase (compensated). The detection method consists of a global threshold of the information computed in the recorded plane ($z = 0$ µm). For this reason, it is called a 2D method.

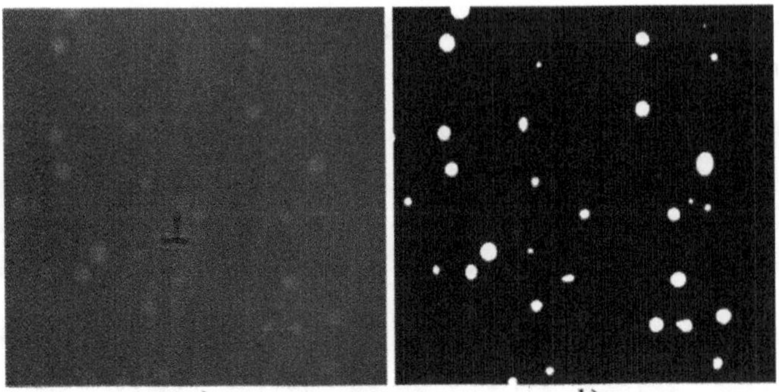

a) b)

Figure 4.6. *a) Compensated phase image of phospholipid vesicles, b) segmented image using the Otsu method*

Depending on the species under study, this global threshold can be obtained with intensity or phase information. For opaque particles, the detection is made on the intensity image, while for transparent particles it is performed on the phase images. Through a global threshold, the detection is performed by analyzing the fluctuations (in phase or in intensity) of the object with respect to the background. Different thresholding methods are available in the literature [SEZ 04], and have to be chosen depending on the image quality. Among those methods, the Otsu method [OTS 79] is a robust thresholding procedure that chooses the threshold level to minimize the intra-class variance of the black and white pixels. This threshold level is then used to convert the gray-scale image (intensity or phase) into a binary image. This process is illustrated in Figure 4.6.

Once the segmentation is performed, the *x-y* coordinates of each particle are determined by computing the centroid of each detected object. Depending on the quality of the images to be segmented, a quantity of non-desirable objects can be detected. These objects are small particles that are not significant, for example, bacteria, dust, bubbles or very small debris. A filtering process is therefore implemented through a selection in the range of area and sizes (major and minor axis) of the detecting objects. The binary image is then cleaned and only contains the particles of interest.

4.4.3.1.2. Detection results

The 2D automated detection method has been applied on various types of algal species as presented in Figure 4.5 as well as on synthetic spherical particle of Figure 4.4.

For the opaque spheres, the detection through the global threshold has been performed on the intensity image of Figure 4.4. The detection of these opaque particles is a simple case and has been performed to test the developed detection procedure. As can be seen in the intensity image, the contrast of the opaque particle is very high compared to the background, which makes it easy to perform the segmentation and then the detection. Approximately 100 holograms with these synthetic particles have been processed with the developed algorithm, giving a mean detection score of 98.4%, with a corresponding standard deviation of 1.3%. This mean rate has been computed by comparing a manual and the automated detection, while the standard deviation has been calculated by repeating five times the detection process. With a global threshold, almost all the particles can be detected directly on the intensity image, as the difference of gray levels between the particle and the background is higher than 180. The contrast also remains high enough for out of focus particles (>150 gray levels) to perform the segmentation easily. Indeed, the only missed particles are those that are close to the border of the holograms. Through this simple preliminary test, the developed detection process has been validated.

The detection software has been applied on alga. Those algae are illustrated in Figures 4.5(a), (b), (c), (g) and (h). For each of these five species, around 1,000 holograms have been processed with the developed detection algorithm. For each species, the set is divided into five subsets of around 200 holograms and the detection process is performed on each subset. Then for comparison, a manual counting process was performed on each subset.

Table 4.1 shows the different mean detection rates, with the corresponding standard deviations. This mean rate has been computed by comparing a manual and an automated count. The other species of Figure 4.5 have been tested with the automated detection method in a qualitative way on a smaller set of holograms to demonstrate the applicability of the technique on a variety of organisms.

Tested organisms	Classical thresholding method	
	μ (%)	σ (%)
Giardia lamblia	89.7	3.4
Chlorella autotrophica	80.4	5.2
Scenedesmus dimorphus	81.3	5.1
Cyanothece sp.	75.2	4.8
Stanieria sp.	62.4	6.1

Table 4.1. *Classical detection method using a global threshold. Mean detection (μ) with the corresponding standard deviation (σ)*

In Table 4.1, we can see a high dispersion in the detection rate for the different organisms. There is, for example, a difference of more than 25% between the detection rate of *Giardia lamblia* (*G. lamblia*) and *Stanieria* sp.

To understand the origin of this large difference, let us observe an example of a *Stanieria* cell, which is the species with the lowest detection score. Figures 4.7(a) and (b) represent the intensity and the phase (compensated) of two *Stanieria* organisms in the microchannel, respectively. The global thresholding process is performed on the phase image, and it results in the binary image as shown in Figure 4.7(c). As seen on the segmented image, the organism at the bottom is missed when applying a global threshold.

a) b) c)

Figure 4.7. *Two Stanieria cells in the recorded plane (0 µm). a) Intensity image, b) phase image, c) global threshold where the organism at the bottom is missed*

By investigating the different reconstruction planes, we can find out why a classical thresholding technique cannot detect the two organisms. Figure 4.8 represents the cropped phase of the two organisms in the different planes of a 100 µm microchannel depth, from –50 µm to 50 µm, obtained by a numerical scanning along the optical axis. The reconstruction of the particle on the top is illustrated in Figures 4.8(a)–(l), while the particle on the bottom is illustrated in Figures 4.8(m)–(x). We can see that the focus plane of the top particle is at 10 µm from the recorded plane and the other particle is refocused at 35 µm.

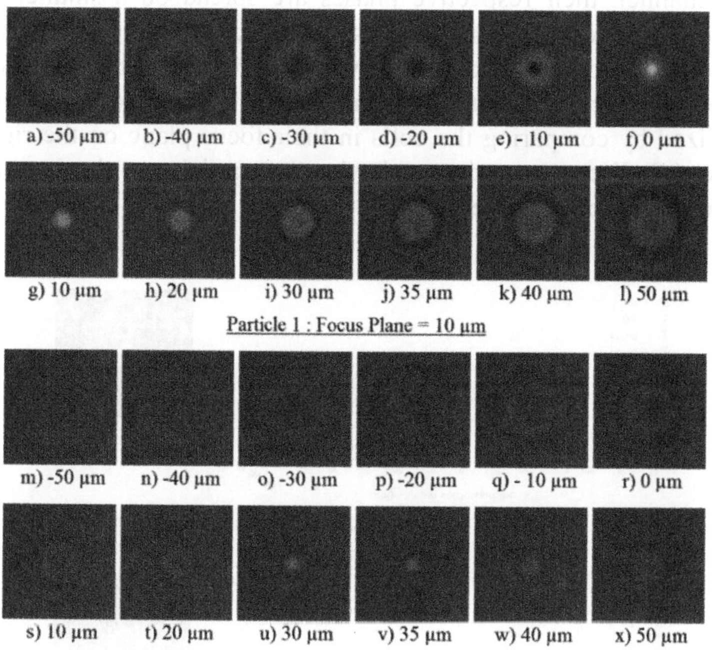

Figure 4.8. *Sequence of the cropped phase of the two organisms of Figure 4.7. The numerical reconstruction is performed from –50 µm to 50 µm*

By analyzing these two propagation sequences, we can observe that as the distance varies, the diffraction becomes more prominent. As a result, the contrast of the object is reduced. This spreading depends on the optical phase variation of the object. For the two *Stanieria* cells in Figure 4.7, their phases are rapidly "diluted" when propagating along the optical axis. The evolution of the optical phase variation as a function of the reconstruction distance for a linear profile is analyzed in Figure 4.9 for the two organisms.

Through the two graphs, we can observe that the spreading and the dilution of the phase quickly become important as the reconstruction distance is varying. The curves in red indicate the focus planes of the two cells. From these focus planes, we observe a clear diminution of the optical phase after a reconstruction distance of 10 µm. And even if the two organisms are just 25 µm away from each other in the microchannel, their respective phases are spread on a smaller scale (10 µm). Moreover, inside the same species, the organisms can have different optical thicknesses, which is related to the variation of the refractive index and the real thickness of the object. This can be visualized by comparing the cells in their focus plane on the right of Figure 4.9. We now understand why the particle on the bottom of Figure 4.7 has been missed.

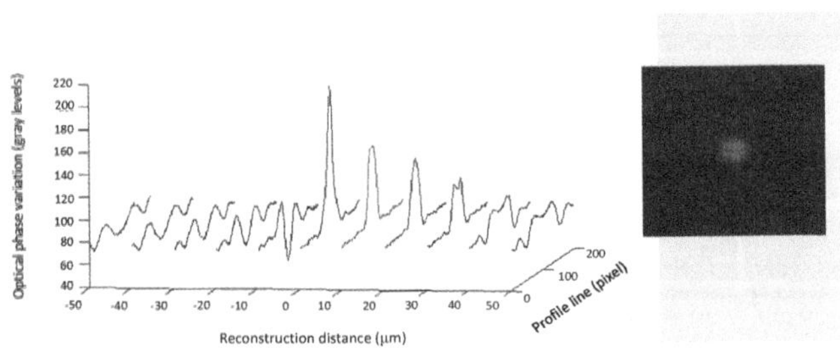

(a)

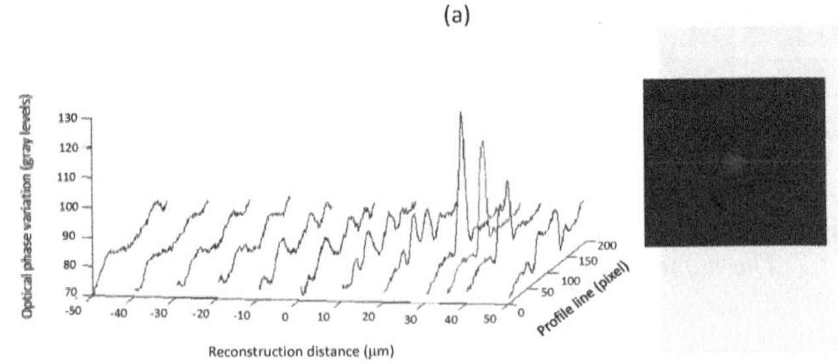

(b)

Figure 4.9. *a) Evolution of the optical phase variation as a function of the reconstruction distance for the top cell of Figure 4.7 along the profile in red. The cell in the right is illustrated in the focus plane. b) It shows a similar evolution in optical thickness for the bottom cell of Figure 4.7. For a color version of this figure, see www.iste.co.uk/picart/digiholography.zip*

The same propagation analysis is then performed on a *Giardia* organism, which is the species with the highest detection rate in Table 4.1. Figure 4.10 illustrates a sequence of different propagating planes. For this organism, we notice that the optical phase variations are small compared to the optical phase variations of *Stanieria*. As the reconstruction distance varies, the difference between the phase of the object and that of the background remains sufficient to detect the organism with a global threshold in the presence of other *Giardia* having different focus planes. This explains the difference of detection rate based on a global threshold method for *Stanieria* and *Giardia*.

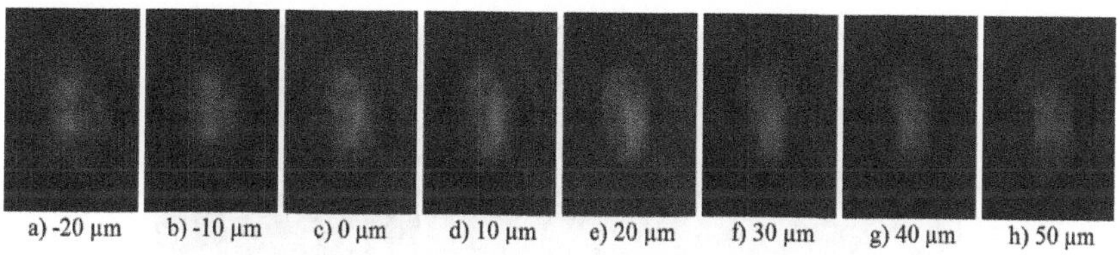

Figure 4.10. *Sequence of a Giardia cyst in the different planes of the microchannel (cropped compensated phase)*

After this analysis, we conclude that a detection based on a global threshold is very easy to implement but not always adequate for organisms recorded in flux inside an experimental volume. Indeed, inside the microchannel, the organism can be recorded at very different focus depths and can thus lead to different optical phase fluctuations inside the same phase image. As such, a classical segmentation method is not sufficient as it just takes into account the information contained in the recorded plane. Moreover, as demonstrated previously, a classical thresholding method depends strongly on the nature of the species, i.e. its optical phase. It can give very different detection rates for different organisms.

For this reason, we investigate a new detection method that takes advantage of the full interferometric information. Our previous analysis leads us to the idea that the reconstruction distance is an important factor and has to be taken into account during the detection. A new procedure that exploits the accumulation of information during the numerical propagation has been developed and is presented in the next section.

4.4.3.2. *Computation of propagating matrices*

To overcome the limitation of the previous classical thresholding method, a new robust method for the 3D automated detection based on the propagation operator is investigated. The method will be illustrated on concentrated samples of *Chlorella* cells in Figure 4.11. We can see in Figure 4.11(b) that some algae (with the arrows) cannot be detected using a global threshold as seen in Figure 4.11(c).

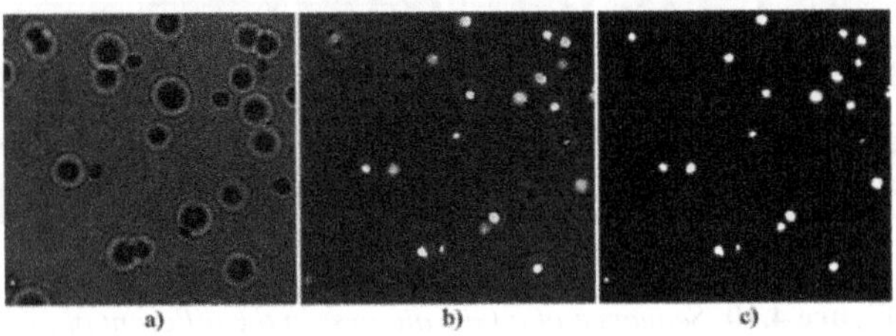

Figure 4.11. *A set of Chlorella alga flowing in a microchannel. a) Intensity image, b) phase image, c) global threshold where some organisms (with the arrow in (b)) are missed. For a color version of this figure, see www.iste.co.uk/picart/digiholography.zip*

4.4.3.2.1. Description of the method

Consider the complex amplitude distribution $u_d(x',y')$ refocused at a distance d along the optical axis by computing the Kirchhoff–Fresnel propagation integral in the paraxial approximation:

$$u_d(x',y') = \exp(jkd) F^{-1}_{x',y'} \exp\left(\frac{-jkd\lambda^2}{2} (v_x^2 + v_y^2) \right) F^{+1}_{v_x,v_y} u_0(x,y) \quad [4.8]$$

where $u_0(x,y)$ is the complex optical field in the recorded plane, $u_d(x',y')$ is the complex amplitude field propagated at a distance d along the optical axis, $k = 2\pi/\lambda$, (x,y) and (x',y') are the spatial variables, respectively, in the focus plane and reconstructed plane, (v_x, v_y) are the spatial frequencies, and $j = \sqrt{-1}$ and $F^{\pm 1}$ denote the direct and inverse 2D continuous Fourier transformations. Equation [4.8] is implemented in a discrete form:

$$u_d(s'\Delta,t'\Delta)=\exp(jkd)F^{-1}_{s',t'}\exp\left(\frac{-jkd\lambda^2}{2N^2\Delta^2}(U^2+V^2)\right)F^{+1}_{U^2,V^2}u_0(s\Delta,t\Delta) \qquad [4.9]$$

where N is the number of pixels in both directions (to match the fast Fourier transform computations), s, t, s', t', U and V are integer numbers varying from 0 to $N-1$, and $F^{\pm 1}$ is now the direct and inverse discrete Fourier transform (DFTs).

The complex amplitude $u_d(s'\Delta,t'\Delta)$ is computed for each refocused distance d inside the experimental volume, from the lower to the upper walls of the microchannel. Then we compute two 3D matrices containing, respectively, the intensity I and the optical phase P for all the refocused distances:

$$\begin{aligned}
I(s'\Delta,t'\Delta,d)&=\mathrm{Re}\left(u_d(s'\Delta,t'\Delta)\right)^2+\mathrm{Im}\left(u_d(s'\Delta,t'\Delta)\right)^2\\
P(s'\Delta,t'\Delta,d)&=\tan^{-1}\left(\frac{\mathrm{Im}\left(u_d(s'\Delta,t'\Delta)\right)}{\mathrm{Re}\left(u_d(s'\Delta,t'\Delta)\right)}\right)
\end{aligned} \qquad [4.10]$$

where Re and Im are the real and the imaginary parts, respectively. Note here that each plane corresponding to a distance d of the 3D P-matrix can be compensated in order to remove the background.

Four matrices can now be computed for, respectively, the minimum and the maximum along the d dimension of the 3D matrices I and P. For these two 3D matrices, we investigate all the d planes and we keep the minimum and the maximum values giving four new matrices:

$$\begin{aligned}
I_{\min}(s'\Delta,t'\Delta)&=\min_d\left(I(s'\Delta,t'\Delta,d)\right)\\
I_{\max}(s'\Delta,t'\Delta)&=\max_d\left(I(s'\Delta,t'\Delta,d)\right)\\
P_{\min}(s'\Delta,t'\Delta)&=\min_d\left(P(s'\Delta,t'\Delta,d)\right)\\
P_{\max}(s'\Delta,t'\Delta)&=\max_d\left(P(s'\Delta,t'\Delta,d)\right)
\end{aligned} \qquad [4.11]$$

where $\min_d$ and $\max_d$ mean, respectively, the minimum and the maximum along all the d planes. It gives rise to four bi-dimensional matrices that can be used simultaneously for the 3D detection of particles. Figures 4.12(a)–(d) illustrate the four matrices of equation [4.11] based on the intensity and phase image of Figure 4.11.

Each of these four matrices can then be automatically thresholded using the Otsu method providing four thresholding parameters, $T_{I_{\min}}, T_{I_{\max}}, T_{P_{\min}}$ and $T_{P_{\max}}$. On the basis of the four thresholded images, a simple logical operation (logical AND) provides the binary matrix illustrated in Figure 4.12(e) and a robust 2D detection becomes possible. In the next section, the performance of this detection method is quantified.

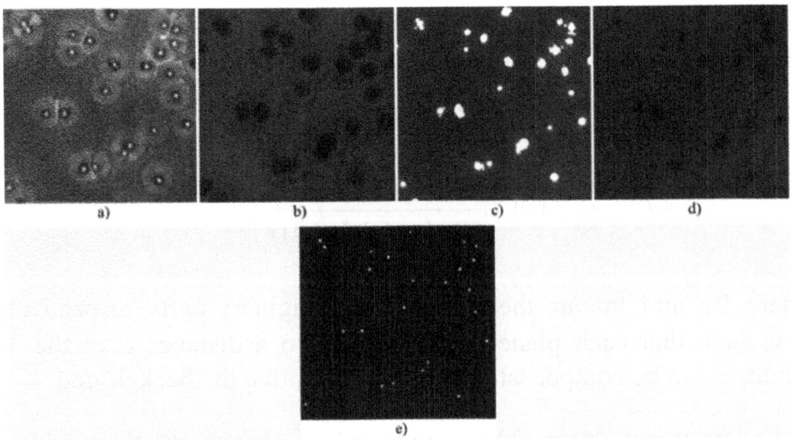

Figure 4.12. *Propagating matrices a) I_{max}, b) I_{min}, c) P_{max} (+compensation) d) P_{min} (+compensation) and e) combination of the four propagating matrices to obtain a binary matrix on which a robust 2D detection can be done*

4.4.3.2.2. Detection results

Applying the same methodology of the previous classical detection method, Table 4.2 exposes the mean detection rate for each of the five studied organisms, with the corresponding standard deviation. Thanks to propagating matrices that easily detect out-of-focus objects, high detection rates are obtained for all the organisms.

Tested organisms	Propagating matrices method	
	μ (%)	σ (%)
Giardia lamblia	96.4	2.2
Chlorella autotrophica	95.1	2.3
Scenedesmus dimorphus	89.4	3.1
Cyanothece sp.	91.4	2.8
Stanieria sp.	92.1	3.1

Table 4.2. *Robust propagating matrices method. Mean detection (μ) with the corresponding standard deviation (σ)*

4.4.3.3. *Comparison of the two methods*

Figure 4.13 compares the detection scores obtained by the classical thresholding method and the propagation matrices method. The improvement of the detection rate due to the newly developed method is clearly demonstrated. *Stanieria* was the species with the lowest detection rate by the first method, and we improve the detection by 30% due to the propagating matrices. We also observed that the new method is considerably less sensitive to the nature of the species, as all the obtained detection rates are very close. The combined detection rate for the five studied organisms is about 93%.

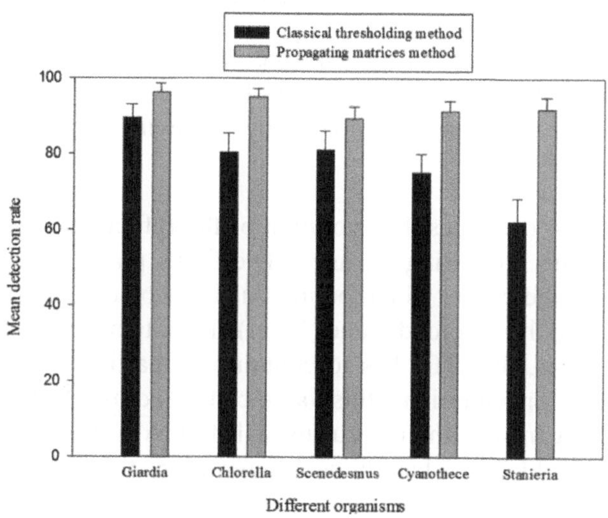

Figure 4.13. *Comparison between the two detection methods: classical thresholding and the method based on propagation matrices method*

4.4.4. *In-depth detection*

Once the particle is detected in the *x-y* plane, the *z* position of the particle has to be determined. This *z* position is obtained by numerically investigating the microchannel volume (refocusing of the different slices). To determine the optimal focus plane, the refocusing criterion based on the integrated amplitude modulus, described in section 4.3, is used. It has been shown that this refocusing criterion is maximal for phase objects as it is the case with the used species. The robustness of this criterion has been demonstrated in [ELM 11], which justifies its use in automated detection.

The refocusing criterion is computed inside a ROI that is established around the detected particle [ANT 08]. This ROI is determined on the basis of the enlarged segmented area of each object in such a way that the diffraction patterns created by the object are also involved in the ROI. The bounding box surrounding the particle is computed and enlarged by 25% on each side. This value has been chosen after a series of tests. After propagating the different organisms, we found that a ROI enlarged by 25% is sufficient to contain the diffraction pattern of the furthest reconstruction plane (50 μm and –50 μm) for a 100 μm microchannel depth.

It has to be emphasized that for the propagating matrices method, the numerical propagation has already been performed and therefore, the refocusing criterion is computed by retrieving the information in the 3D matrices.

Figure 4.14 describes the evolution of the refocusing criterion as a function of the refocusing distance. The focus plane is given by the maximum of the curve, which permits us to refocus a *G. lamblia* cyst at 18 μm from the recorded plane (0 μm). To obtain a correct value for the focus plane, this ROI has to contain one particle. In the case of high particle concentrations, this value can be adjusted. In the case of clusters of organisms, the refocusing will fail. A particular processing has been developed to handle such aggregates through a separation procedure [ELM 13].

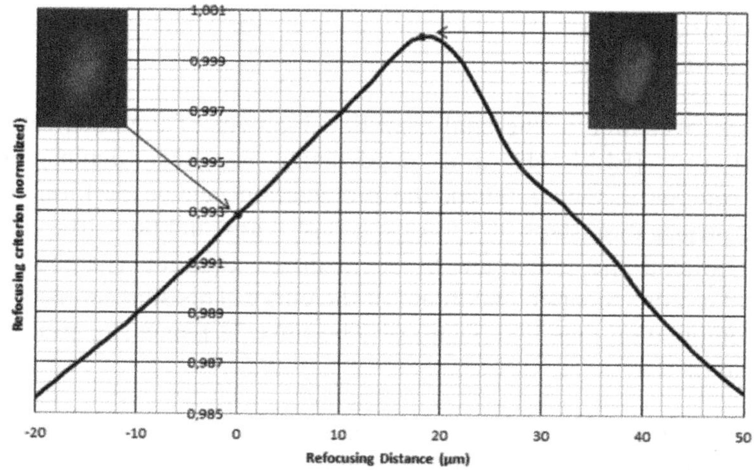

Figure 4.14. *Evolution of the refocusing criterion as a function of the refocusing distance. The species on the left is at the recorded plane of 0 µm. The focus plane of the organism is given by the maximum of the curve, and is at 18 µm as illustrated by the image on the right*

Refocus scanning is implemented to optimize the computation time. To refocus a set of *n* particles in a hologram, a usual method is to compute for the whole hologram the propagation equation for the set of planes and to save every plane information (intensity and phase information) in a big 3D matrix (2D for the pixel dimension (vertical and horizontal axes) and *z* for the refocusing distance). Once a particle is detected, the information needed is recovered in this matrix inside a ROI binding each particle. The refocusing criterion is computed inside each ROI to obtain the *z* position of the particle. To refocus a whole hologram of 1,024 × 1,024 pixels, the computation time for 101 planes (from −50 µm to 50 µm) is 34 s, (this is done on an i7-870 Intel Core Processor 2.93 GHz). This computation time is largely due to the two fast Fourier transforms (FFTs) of the reconstruction equation. To decrease the computation time, we perform the propagation sequentially inside the enlarged ROI of each particle instead of doing it on the whole hologram. For a 128 × 128 pixels ROI, the obtained computation time for 101 planes is 0.45 s. Then, if the number of particles is lower than 75, it is faster to propagate locally inside each ROI instead of propagating the whole hologram. In practice, the number of particles does not exceed 30, which validates this optimization approach.

Moreover, due to the analytical development of the used refocus criterion in [ELM 11], a ε^2 dependency has been demonstrated (ε = refocus distance). Then, the propagation can be performed by using propagation steps of 5 μm instead of 1 μm, and the curve can be fitted around the maximum by a quadratic equation (parabola) to extract a more accurate maximum of the curve. This step can decrease the computation time by a factor of 5 because 21 planes are computed instead of 101.

When the *z* position has been established, the particle is re-segmented in its focus plane to obtain a precise delimitation of the object. Indeed, the first segmentation (performed in the out of focus recorded plane) only provides a rough estimation of the size of the object. This first estimation is sufficient to determine the ROI for the z scanning but cannot provide any precise information of the object for a future classification. By segmenting the object in the focus plane, the precise shape can be extracted and a more precise *x-y* position of the object can be computed. Thereby, the 3D position of each object is obtained.

4.4.5. *Discussion*

Two 3D detection methods have been presented in the previous section. A usual method has been investigated and is based on a global threshold that detects the organisms on the intensity or the phase image, depending on the nature of the species. This method is commonly used to automatically detect objects in fluxes. It has been demonstrated that this classical method fails when objects of low optical thickness are far away from the recorded plane.

To overcome this limitation, a new robust method has been proposed that takes advantage of the full interferometric information contained in a digital hologram. This method is based on the propagation of the complex amplitude information resulting in the computation of four matrices. Robust 3D detection can therefore be achieved. The proposed detection method has been successfully applied on a variety of species, giving high detection rates (up to 95%).

However, the disadvantage of this new method compared to the classical method is the computation time. Indeed, to create the four matrices, the propagation has to be done on the whole hologram (1,024 × 1,024 pixels) and the process takes 34 s. This needs to be done even if there is only one object in the field-of-view (FOV). Indeed, the *x*-*y* detection is performed on the basis of the four matrices obtained after a whole propagation of the hologram.

Using the classical method, we have developed a local propagation (in a ROI of 128 × 128 pixels) that takes 0.45 s per organism. This local scanning inside a ROI is advantageous compared to a whole propagation if the number of particles is lower than 75. Using the new robust method, a whole scanning needs to be performed for each hologram before the detection, even if there is only a few organisms.

The usual method can be used for organisms with a high refractive index, or with an optical phase that varies slightly along the optical axis. It has been shown that for *Giardia*, the usual method gives an acceptable detection rate (~90%). In our laboratory, this usual method is currently used for objects such as vesicles and RBCs due to the high refractive index of such organisms. For the other investigated objects exposed in this section, a global threshold can miss up to 40% and the new robust method has to be used.

The price to pay is the increased processing time which can be critical for real-time applications for example. Note here that our algorithms are built as a first step in terms of developments. No optimization effort has been attempted to decrease the computation time. The continuous evolution of the performances of computer processors facilitating parallel computing as well as the implementation on graphics processing unit (GPU) can very easily overcome this drawback. In this chapter, we consider that the efficiency of the developed algorithm is of bigger importance than the optimization of the computation time.

In the next section, we illustrate an application of the 3D detection process by analyzing the automated monitoring of RBC's dynamics.

4.5. Application

Blood is a complex biological fluid made of more than 50% of RBCs. They are the principal means of delivering oxygen to the body tissues via the blood flow through the circulatory system. They take up oxygen in the lungs or gills and release it while squeezing through the body's capillaries. These cells' cytoplasm is rich in hemoglobin, an iron-containing biomolecule that can bind oxygen and is responsible for the blood's red color. RBCs have a disk diameter of approximately 6.2–8.2 μm, a thickness at the thickest point of 2–2.5 μm and a minimum thickness in the center of 0.8–1 μm. They are much smaller than most of the other human cells. They are highly deformable and can adopt various shapes depending on the channel they are flowing in.

The distribution of RBCs in a confined flow is inhomogeneous and shows a significant depletion near the walls due to a competition between migration away from the walls and shear-induced diffusion resulting from interactions between particles [GRA 13].

The migration of blood cells forms the physical basis of the formation of a cell-free plasma layer near vessel walls in the microcirculation. Furthermore, with the very high concentration of RBCs in the human blood, strong hydrodynamic interactions take place. Consequently, from a rheological point of view, blood is a complex fluid leading to complex flow patterns in the microcirculation networks where the diameter of blood vessels becomes comparable to the cell size. Those inhomogeneities have important consequences on the transport of oxygen and may lead to situations where one branch of a capillary bifurcation is weak or not irrigated by RBCs.

In this section, we report the investigation and quantification of the lift force that pushes cells away from the vessel walls when a shear flow is applied. The experimental cell is made up of two circular glass plates with a gap of 180 µm. The bottom disk is fixed while the top disk is rotating which creates a shear flow inside the experimental cell. The fluid is injected via an inlet in the center of the bottom glass plate. To cope with the DHM requirements in terms of concentration and limit the hydrodynamic interactions between pairs of cells, we work with highly diluted and washed blood. The resulting fluid is made of RBCs in an external solution (bovine serum albumin (BSA) and phosphate buffered saline (PBS)) alone or in combination with Dextran (typically 1%). By adding Dextran, we can increase the viscosity of the external fluid, and easily control the lift velocity. To avoid any screening of the lift by sedimentation, we performed these experiments in microgravity during parabolic flight campaigns onboard the Airbus A300 Zero-G from Novespace. Each microgravity period lasts 20 s and is monitored by DHM at a frequency of 24 holograms per second. The processing of the holograms is fully automated and, depending on the concentration of the sample, for each sequence, 10,000–35,000 cells are detected by the classical thresholding method and analyzed collecting reliable statistics.

The lift force of deformable objects was the subject of a theoretical study by Olla [OLL 97] and a predictive law has been determined which gives the distance from object to wall as a function of the shear rate (γ), the radius of the object (*R*), the time (*t*) and a parameter (*U*) describing the lift velocity and containing the viscosity ratio (between inner and outer fluid). The law can be reduced to the following expression:

$$\frac{\langle Z^3 \rangle}{R^3} = 3U\gamma t \qquad [4.12]$$

which exhibits a linear dependency between the rescaled mean position of objects to the power of 3 and the shear rate multiplied by time. Experimental results confirm this predictive law very well by exhibiting linear curves with different lift velocities for the different viscosity ratios (Figure 4.15) [CAL 08]. By symmetry, a tumbling rigid object should not migrate on average. The non-zero lift suggests that RBCs deformability leads to symmetry breaking: it is stretched when oriented in the direction of the elongational component of the flow, while it is compressed when orthogonal, resulting in an averaged asymmetric shape, leading to a migration law similar to the law known for a fixed shape and orientation. By increasing the external viscosity, the stresses on the RBC membrane are higher and lead to increased deformation, which, in turn, enhances the lift as clearly depicted in Figure 4.15.

These results demonstrate clearly the capability of digital holographic microscopy to monitor RBCs suspension in micro-flow circulation, which is a step toward the better understanding of blood circulation.

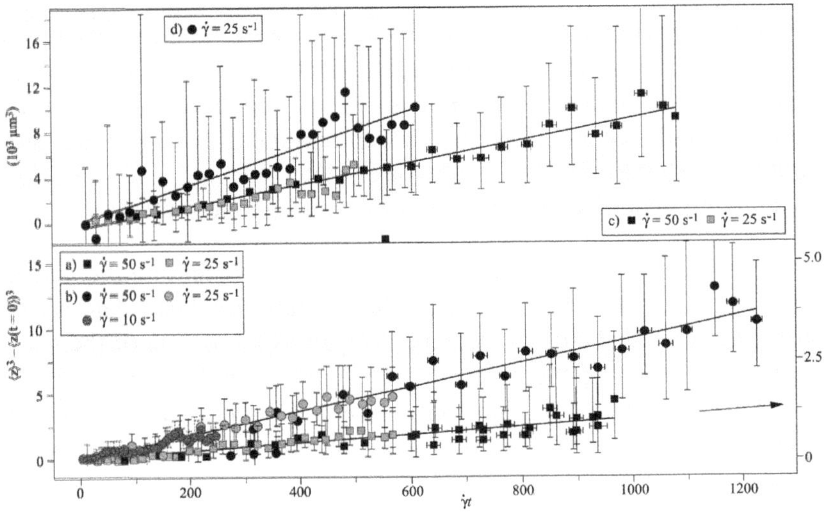

Figure 4.15. *RBC – wall distance <z3> vs. γt for different outer viscosities: a) 1.4 mPa.s; b) 6.1 mPa.s; c) 9.3 mPa.s; d) 13 mPa.s. Full lines indicate fit to Olla's prediction, with UR3 = 0.36, 3.1, 3.2, 5.4 μm^3, respectively*

4.6. Conclusions

Object detection procedures are an essential step for any real application. In this chapter, after having described specific tools for DHM, we have presented two methods to automatically achieve the 3D positions of particles moving in an experimental cell. The first detection method is based on a global threshold. This method is easy to implement and is not time-consuming. It gives an acceptable detection rate for the species we tested (combined detection rate = 77.8%). However, it has been demonstrated that this method strongly depends on the species' nature as a global threshold fails for some species whose optical phases are rapidly spread by propagating along the optical phase. These organisms recorded far from their focus plane will therefore be missed by a global threshold.

To overcome this drawback, we have developed a robust detection procedure that takes advantage of the full interferometric information available in digital holography. This method is based on the propagation of the complex amplitude information resulting in the computation of four new matrices that accumulate the information along the propagation axis. By combining these matrices, a robust *x*-*y* detection can be achieved. The combined detection rate for the species we tested is about 93% that considerably increased the amount of detected particles.

The developed automated 3D detection procedures have been used for different applications. In this chapter, we applied our detection method to automatically analyze the dynamics of RBCs in order to better understand blood circulation.

4.7. Bibliography

[ANT 08] ANTKOWIAK M., CALLENS N., YOURASSOWSKY C., *et al.*, "Extended focused imaging of a microparticle field with digital holographic microscopy", *Optics Letters*, vol. 33, pp. 1626–1628, 2008.

[CAL 08] CALLENS N., MINETTI C., COUPIER G., *et al.*, "Hydrodynamical lift of vesicles under shear flow in microgravity", *Europhysics Letters*, vol. 83, p. 24002, 2008.

[DCA 14] DI CAPRIO G., EL MALLAHI G., FERRARO P., *et al.*, "4D tracking of clinical seminal samples for quantitative characterization of motility parameters", *Biomedical Optics Express*, vol. 5, pp. 690–700, 2014.

[DOH 14] DOHET-ERALY J., YOURASSOWSKY C., DUBOIS F., "Refocusing based on amplitude analysis in color digital holographic microscopy", *Optics Express*, vol. 39, pp. 1109–1112, 2014.

[DUB 99] DUBOIS F., JOANNES L., LEGROS J.C., "Improved three-dimensional imaging with a digital holography microscope with a source of partial spatial coherence", *Applied Optics*, vol. 38, pp. 7085–709, 1999.

[DUB 02] DUBOIS F., MONNOM O., YOURASSOWSKY C., *et al.*, "Border processing in digital holography by extension of the digital hologram and reduction of the higher frequencies", *Applied Optics*, vol. 41, pp. 2621–2626, 2002.

[DUB 04] DUBOIS F., NOVELLA REQUENA M.-L., MINETTI C., *et al.*, "Partial coherence effects in digital holographic microscopy with a laser source", *Applied Optics*, vol. 43, pp. 1131–1139, 2004.

[DUB 06a] DUBOIS F., SCHOCKAERT C., CALLENS N., *et al.*, "Focus plane detection criteria in digital holography microscopy by amplitude analysis", *Optics Express*, vol. 14, pp. 5895–5908, 2006.

[DUB 06b] DUBOIS F., CALLENS N., YOURASSOWSKY C., *et al.*, "Digital holographic microscopy with reduced spatial coherence for 3D particle flow analysis", *Applied Optics*, vol. 45, pp. 964–961, 2006.

[ELM 11] EL MALLAHI A., DUBOIS F., "Dependency and precision of the refocusing criterion based on amplitude analysis in digital holographic microscopy", *Optics Express*, vol. 19, pp. 6684–6698, 2011.

[ELM 13a] EL MALLAHI A., MINETTI C., DUBOIS F., "Automated three-dimensional detection and classification of living organisms using digital holographic microscopy with partial spatial coherence source: application to the monitoring of drinking water resources", *Applied Optics*, vol. 52, pp. A62–A80, 2013.

[ELM 13b] EL MALLAHI A., DUBOIS F., "Separation of overlapped particles in digital holographic microscopy", *Optics Express*, vol. 21, pp. 6466–6479, 2013.

[GOO 05] GOODMAN J.W., *Introduction to Fourier Optics*, McGraw-Hill, 2005.

[GRA 13] GRANDCHAMP X., COUPIER G., SRIVASTAV A., *et al.*, "Lift and down-gradient shear-induced diffusion in red blood cell suspensions", *Physical Review Letters*, vol. 110, p. 108101, 2013.

[KRE 86] KREIS T., "Digital holographic interference-phase measurement using the Fourier-transform method", *Journal of the Optical Society of America A*, vol. 3, pp. 847–855, 1986.

[LAN 08] LANGEHANENBERG P., KEMPER B., DIRKEN D., *et al.*, "Autofocusing in digital holographic phase contrast microscopy on pure phase objects for live cell imaging", *Applied Optics*, vol. 47, pp. D176–D182, 2008.

[MAR 05] MARQUET P., RAPPAZ B., MAGISTRETTI P.J., *et al.*, "Digital holographic microscopy: a non-invasive contrast imaging technique allowing quantitative visualization of living cells with subwavelength axial accuracy", Optics Letters, vol. 30, pp. 468–470, 2005.

[MIN 08] MINETTI C., CALLENS N., COUPIER G., *et al.*, "Fast measurement of concentration profiles inside deformable objects in microflows with reduced spatial coherence digital holography", *Applied Optics*, vol. 47, pp. 5305–5314, 2008.

[OLL 97] OLLA P., "The lift on a tank-treading ellipsoidal cell in a shear flow", *Journal of Physics II*, vol. 7, p. 1533, 1997.

[OTS 79] OTSU N., "A threshold selection method from gray-level histograms", *IEEE Transactions on Systems, Man, and Cybernetics*, vol. 9, pp. 62–66, 1979.

[SEZ 04] SEZGIN M., SANKUR B., "Survey over image thresholding techniques and quantitative performance evaluation", *Journal of Electronic Imaging*, vol. 13, pp. 146–165, 2004.

[SCH 02] SCHNARS U., JÜPTNER W., "Digital recording and numerical reconstruction of holograms", *Measurement Science and Technology*, vol. 13, pp. 85–101, 2002.

[YOU 14] YOURASSOWSKY C., DUBOIS F., "High throughput holographic imaging-in-flow for the analysis of a wide plankton size range", *Optics Express*, vol. 22, pp. 6661–6673, 2014.

5

Quantitative Phase-Digital Holographic Microscopy: a New Modality for Live Cell Imaging

Quantitative phase microscopy (QPM) has recently emerged as a new powerful quantitative imaging technique well-suited to non-invasively explore, with a nanometric axial sensitivity, transparent specimens. Among several novel optical QPM techniques, quantitative phase digital holographic microscopy (QP-DHM), as described in Chapter 1, is a particularly powerful approach to explore cell structure and dynamics in a label-free manner. In section 5.1 of this chapter, the most relevant QPM and especially QP-DHM applications in the field of cell biology are summarized. A particular emphasis is put on the original biological information which can be derived from the QP signal (QPS). In section 5.2, recent promising applications obtained in particular with QP-DHM in the field of high content screening (HCS) are presented. In section 5.3, the important issue concerning the development of multimodal microscopy is discussed and illustrated through concrete examples, including combination of QP-DHM with fluorescence microscopy, Raman spectroscopy (RS) and electrophysiology. Finally, section 5.4 describes how important questions related to the field of

---

Chapter written by Pierre MARQUET, Benjamin RAPPAZ and Nicolas PAVILLON.

cellular neuroscience, namely the possibility to optically resolve neuronal network activity and spine dynamics, can be discussed with both QP-DHM in a multimodality approach and QP tomography (QPT).

5.1. Introduction

Elucidating the mechanisms underlying physiological and pathophysiological cellular processes requires the possibility to quantitatively and non-invasively observe cell structure and dynamics down to the cellular and subcellular levels [KUM 14]. Considering that most biological cells are transparent, i.e. they differ only slightly from their surroundings in terms of optical properties (including absorbance, reflectance, etc.), it results that obtaining a comprehensive understanding of physiological and pathophysiological cellular processes with classical optical approaches (bright-field) remains a difficult challenge. Consequently, a wide range of advanced optical approaches has been developed in cell biology [INO 97]. Specifically, the development of both new modes of contrast generation and new forms of high-resolution microcopy techniques allowing us to break the so-called "diffraction limit" of resolution, which limits the smallest observable details in the image to approximately 200–500 nm, is a very active area of research in cell biology. Accordingly, fluorescence microscopy [HAN 13] is today a powerful and widely used cell imaging technique [COR 14].

Taking into account the non-invasiveness requirement, contrast generating modes based on intrinsic cell properties are particularly appealing. Among them, those operating on the wave front phase information, which have already been developed in the mid-20th Century, have demonstrated their relevance for non-invasive visualization of cell structure, in particular due to the Zernicke's invention of phase contrast (PhC) [ZER 42] and the DIC techniques developed by Smith [MIT 56] and Nomarski [NOM 50].

Otherwise, the past few decades witnessed the emergence of several novel optical QPM techniques, quite simple to implement and based on interferometric or holographic approaches including DHM which have consistently started to become relevant imaging techniques for exploring cell structure and dynamics [POP 08, LEE 13, MAR 13].

Accordingly, in section 5.1, after having explained the origin of the QPS and described the wealth of biological information it contains, we briefly discuss some of the most relevant applications related to cell biology provided by different QPM techniques. Special attention is given to the applications which can be specifically discussed by QP-DHM due, in particular, to its numerical extended depth of focus and autofocusing capabilities resulting, as explained in Chapter 1, from the possibility to propagate the whole object wave (amplitude and phase) diffracted by the observed specimen during the numerical reconstruction of the digitally recorded holograms. Section 5.2 presents the possibilities for efficiently performing label-free HCS with QP-DHM, in particular, due to its capabilities offered by the numerical reconstruction as mentioned above. Section 5.3 describes some multimodality approaches combining QP-DHM with, in particular, fluorescence microscopy, RS and electrophysiology. Indeed, physiological or pathophysiological cell processes, usually being multifaceted, can be efficiently grasped only with techniques providing a view which includes many of their aspects at once. Thus, the development of high-resolution multimodal imaging approaches providing simultaneous measurements of a set of multiple relevant biophysical cell parameters is a prerequisite for dissecting complex cellular processes. In the final section, original results demonstrate how a multimodal approach combining QP-DHM with fluorescence microscopy and electrophysiology can discuss two important issues in neurobiology, namely the optical resolution of local neuronal network activity by performing a simultaneous multiple sites optical recording of transmembrane currents and the dynamic visualization of dendritic spine – the loci of structural and functional synaptic plasticity – due to a QPT approach. Finally, some perspectives are presented. It should be noted that most of the applications presented in this chapter have been performed with DHM in an off-axis configuration, whose main technical aspects as well as those relative to QPT have been described in Chapter 1.

5.2. Cell imaging with quantitative phase DHM

5.2.1. *The origin and content of the quantitative phase signal*

Biological specimens such as living cells and tissues are usually phase objects, that is they are transparent, namely differing from the surroundings only by a slight difference of refractive index (RI), which results from the presence of organic molecules in cells, including proteins, deoxyribonucleic acid (DNA), organelles, and nuclei. These transparent specimens also called phase objects can be made visible, in particular their fine subcellular structural organization, by PhC or DIC as mentioned in section 5.1. Basically, these two non-invasive contrast-generating techniques are based on their capacity of transforming in detectable intensity modulation the minute relative phase shift that a transparent microscopic object induces between the transmitted wave light and the undeviated background wave (PhC) and between two orthogonally polarized transmitted waves (DIC). However, PhC as well as DIC do not allow the direct and quantitative measurement of this phase shift or optical path length. Indeed, PhC suffers from well-known optical artifacts, namely the halo and shading off, making the intensity modulation at any pixel not directly proportional to the phase change due to the corresponding area of the cell. Regarding DIC, the intensity modulation is proportional to the spatial directional gradient along the shear direction of the intracellular RI resulting in a spurious asymmetry in the image. Consequently, DIC or PhC signal variations are only qualitative and remain difficult to interpret in terms of quantitative modification of specific biophysical cell parameters [RAP 14]. However, we might note some attempts to use DIC and PhC as quantitative imaging techniques [COG 97, ISH 99, KIN 08, FU 10, KOU 10, GAO 11, YIN 12]. In contrast, QPM including QP-DHM visualizes cells by quantitatively providing the phase retardation that they induce on the transmitted wave front [MAR 05]. 3D perspective images of such a phase retardation corresponding to a mouse cortical neuron in culture and a living human red blood cell (RBC) are shown in Figure 5.1(a) and (b), respectively.

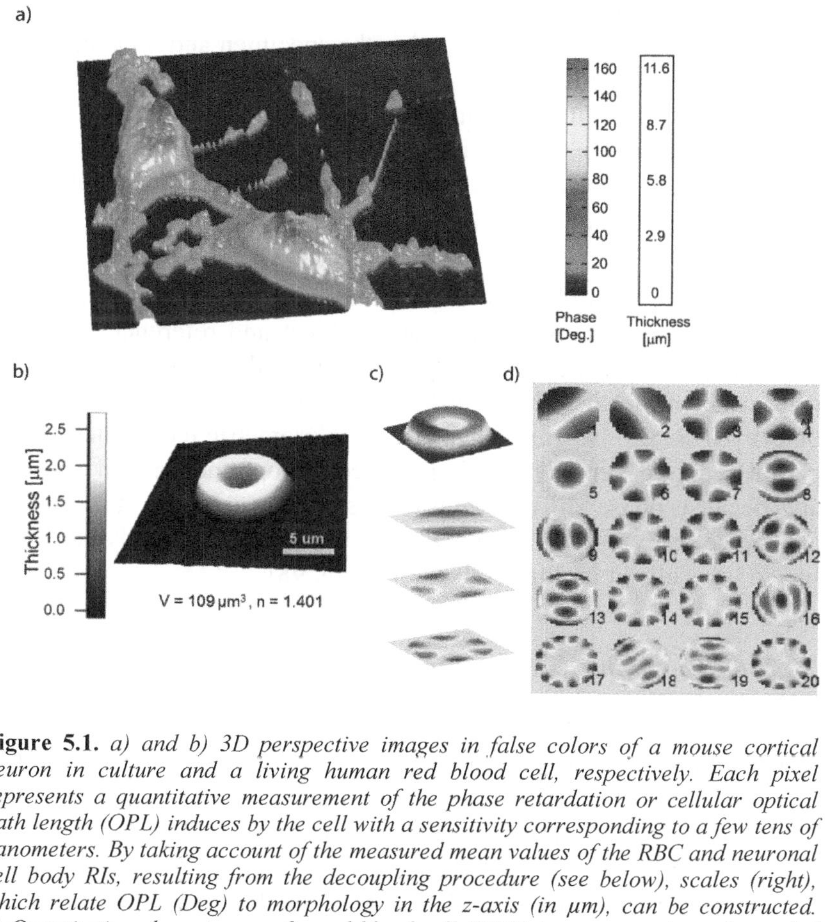

Figure 5.1. *a) and b) 3D perspective images in false colors of a mouse cortical neuron in culture and a living human red blood cell, respectively. Each pixel represents a quantitative measurement of the phase retardation or cellular optical path length (OPL) induces by the cell with a sensitivity corresponding to a few tens of nanometers. By taking account of the measured mean values of the RBC and neuronal cell body RIs, resulting from the decoupling procedure (see below), scales (right), which relate OPL (Deg) to morphology in the z-axis (in µm), can be constructed. c) Quantitative phase image of a red blood cell (RBC) in a perspective color-coded representation and projections of three principal component analysis (PCA) modes of the RBC cell membrane fluctuations. d) Degeneracy-free representation of 20 PCA modes, measured experimentally. For a color version of this figure, see www.iste.co.uk/picart/digiholography.zip*

Definitively, the QPS signal is given by the following expression:

$$\varphi = \frac{2\pi}{\lambda}(\overline{n}_c - n_m)d \qquad [5.1]$$

where d is the cellular thickness, $\overline{n}_c$ is the intracellular RI averaged over the OPL of optical rays crossing the specimen and n_m is the RI of the surrounding medium. The cellular QP contrast thus results from the fact that the intracellular RI is different from the RI of the perfusion solution n_m. Equation [5.1] makes it clear that the QPS contains cell information about both the intracellular content related to $\overline{n}_c$ and the cell morphology related to the thickness d.

Finally, it should also be mentioned that the invention of interference microscopy with separate object and reference beams at the end of 19th Century already provided a direct and quantitative measurement of the phase difference introduced by the observed object [BAR 52]. However, because interferometric techniques at that period required very demanding and costly optomechanical designs, and because phase shifts are very sensitive to experimental artifacts and noise, only a few attempts have been reported to dynamically image live cells in biology [DUN 93, DUN 98].

5.2.2. *Cell counting and classification analysis*

Practically, the QPS is particularly well-suited to the development of algorithms, allowing automated cell counting [MIL 02, MOL 08, SEO 10, MIH 11, BIA 13], recognition and classification [MOO 07, DAN 10, JAV 10a, MOO 10, ANA 11, LIU 11, MEM 11, MOO 12, YI 12, YI 13]. As far as RBCs are concerned, QPM allows us to achieve a morphometry-based RBCs analysis. Indeed, the intracellular content of mature nucleus- and organelle-free RBCs, which can be assimilated, from an optical point of view, to a homogeneous solution containing mainly hemoglobin, induces a QP shift directly proportional to the RBC thickness [POP 05, BER 08, RAP 08,

LIU 11, MIH 11, MIR 11a, MEH 12, MOO 12, SHA 12, YI 12, MOO 13]. Since the QPS is also sensitive to intracellular RI, it provides specifically some information about the intracellular content. Based on this information, interesting applications allowing discrimination between physiological and pathophysiological states have been achieved, particularly in the fields of assisted reproduction [CRH 11, MEM 11, MER 13] and cancer research [MAN 05, JAN 09, WAN 10, WAN11].

5.2.3. *Exploration of cell movements and dynamics*

QPM approaches as non-scanning techniques are usually well-suited to monitor dynamic cell processes. Dynamic motions during the migrations of human dendritic cells [GAR10] and subcellular contraction of embryonic cardiomyocytes [ELL 07, ANS 12] have been successfully measured. On the other hand, QP-DHM allowing us, due to the numerical propagation of the reconstructed object wave front, to apply autofocusing [LIE 04, CHO 09, KIM 09, LAN 09, WAN 09, ZHA 10, TOY 12] and extended depth of focus [FER 05, FER 06, ANT 08a, MCE 08, BER 09, COL 10, PAN 13] has opened the possibility of efficiently monitoring cell migration in 3D [MAN 06, DAN 07, SUN 08, CHO 09, LAN 09, BAE 10, JAV 10b, BOH11, MEM 12, MER 12]. In addition, these numerical possibilities offer an alternative to the shallow depth of field of conventional microscopy, which hampers any fast 3D tracking of cells in their environment, specifically when microfluidic devices are considered [MIC 11, PAT 12]. Coherently, autofocusing and extended depth of focus abilities also facilitate the 3D tracking of micro- or nanoparticles [XU 03, ANT 08b, BAE 10, WAR 10], particularly in combination with approaches capable of second harmonic generation [HSI 09, SHA 10], the use of which is highly promising in medical fields including drug carrier, tumor detection and treatment, and gene therapy.

5.2.4. *Dry mass, cell growth and cell cycle*

As described by equation [5.1], the measured QP shift induced by an observed cell on the transmitted light wave front is proportional to the intracellular RI, which mainly depends on protein content. Therefore, this measure can be used to directly monitor protein production, owing to a relation established more than 50 years ago by Barer [BAR 52]. Practically, the phase shift induced by a cell is related to its dry mass (DM) by the following equation (converted to the International System of Units):

$$DM = \frac{10\lambda}{2\pi\alpha}\int_{S_c} \Delta\varphi ds = \frac{10\lambda}{2\pi\alpha}\overline{\Delta\varphi}S_c \qquad [5.2]$$

where $\overline{\Delta\varphi}$ is the mean phase shift induced by the whole cell, λ is the wavelength of the illuminating light source, S_c is the projected cell surface and α is a constant known as the specific refraction increment (in cubic meters per kilogram) and related to the intracellular content. α is approximated by 1.8×10^{-3}–2.1×10^{-3} m^3 kg^{-1}, when considering a mixture of all the components of a typical cell [BAR 52].

Recently, several groups using various QPM techniques have begun to exploit this phase–DM relationship to study the dynamics of cell growth as well as to the characterization of the cell cycle [ZIC 95, POP 08a, RAP 09, KEM 10, MIR 11b, BYU 12]. As far as RBCs are concerned, this relationship has also been explicitly and implicitly used to calculate hemoglobin content, a relevant clinical parameter [RAP 08, YUS 09, JAN 12, PHA 12, RIN12].

5.2.5. *Cell membrane fluctuations and biomechanical properties*

RBCs, while being carried throughout the vascular tree to deliver oxygen from the lungs to the tissues, are squeezed as they pass through capillaries often smaller than their diameter. This ability can be attributed to the remarkable elastic properties of the membrane structure. As a result of these elastic properties, RBCs show cell membrane fluctuations (CMFs) at the nanometric scale, which is often called flickering. Owing to their high sensitivity allowing us to quantitatively measure RBC membrane fluctuations over the whole cell surface, different QPM techniques have shed new light on these CMFs especially by providing quantitative information about the biomechanical properties of RBC membrane [POP 08a, RAP 09, PAR 10a, LEE 11, PAR 11, JAN 12] in particular by using principal component analysis (PCA) [BOS 12]. We should also mention the integration of QPM with optical tweezers, which represents a very promising tool, especially with respect to monitoring trapped objects along the axial direction including nanoparticles [HIG 11, HAY 12], as well as manipulating and testing biomechanical properties of cells [DAN 10, MAU 10, CAR 11a, CAR 11b, ESS 12, KEM 13].

5.2.6. *Dynamics of absolute cell volume and transmembrane water movements*

Membranes of animal cells are highly permeable to water; movement of water across membranes is therefore dictated in a large part by osmotic pressure gradients [FRI 06, HOF 09]. Any imbalance in intracellular and extracellular osmolarity is thus paralleled by water movements across cell membranes affecting cell volume along with the concentration of intracellular compounds [PAS 10]. However, even at constant extracellular osmolarity, volume constancy of any mammalian cell is continuously challenged by the normal activity of the cells [STR 04]. Thus, the maintenance of a constant volume is critical for cell homeostasis and requires the continued operation of cell volume regulatory mechanisms [FRI 06]. Water crosses cell membranes though several routes (simple diffusion through the lipid bilayer, transmembrane proteins such as specialized water channels, aquaporins, AQP, etc.) [HOF 09]. Consequently, the accurate measurement of both the absolute cell volume dynamic and water membrane permeability P_f represents indices which can provide highly relevant information about the mechanisms involved in the cell volume regulatory processes. However, the non-invasive measurement of these two parameters at the single cell level is not a trivial issue [BOS 13]. Below, we present how these parameters can be calculated from the QPS although, as indicated by equation [5.1], information concerning the intracellular content related to $\overline{n}_c$ is intrinsically mixed with morphological information relating to a thickness d. Practically, different strategies have been developed to separately measure cell morphology and RI. Some authors [LUE 06, KEM 07] measured the intracellular RI by trapping cells between two cover slips separated by a known distance. However, this approach, which prevents cell movement, precludes the possibility of exploring dynamic cellular processes, including cell volume regulation. Recently, spectroscopy phase microscopy approaches [PAR 09, JAN 12, RIN 12] have discussed this limitation, at least as far as cells with high intrinsic dispersion properties are considered, including RBCs (owing to the presence of hemoglobin pigment). However, such spectroscopic approaches are only applicable to a very limited variety of cell types, as most of them have intrinsic dispersion almost identical to that of water. We have developed another approach, called the decoupling procedure, to separately measure the parameters $\bar{n}_c$ and d from the phase signal φ, based on a modification of the extracellular RI n_m. Basically, this method consists of performing a slight alteration of the extracellular RI n_m and recording two holograms corresponding to the two different values of n_m, providing thus two QP images from which $\bar{n}_c$ and d can be calculated. We have considered two different approaches to modify n_m: the first approach requires sequentially perfusing a standard cell perfusion solution and a second solution with a different RI but with the same osmolarity to record the two corresponding holograms, at a single wavelength [RAP 05]. Practically, this procedure has allowed us to quantitatively measure some highly relevant RBC parameters including mean corpuscular volume and mean corpuscular hemoglobin concentration [RAP 08]. However, owing to the solution exchange time, this approach precludes the possibility of monitoring dynamic changes of cell morphology and RI that occur during fast biological processes. To overcome these drawbacks, we have developed a second approach, based on a dual-wavelength DHM (DW-DHM) [KUH 07] which exploits the dispersion of the extracellular medium, enhanced by the utilization of an extracellular dye, to achieve separate measurements of the intracellular RI and the absolute cell volume in real time [RAP 08]. These approaches have been successfully applied to studying the osmotic water membrane permeability P_f – representing the water volume flux per unit of time per unit of membrane surface for a given applied osmotic gradient – by monitoring cell volume changes while retaining the cell functionality [BOS 13]. In addition, resulting from the linear relationship between the intracellular RI and the DM as well as from a DM balance equation, it has been possible to determine the RI of the transmembrane flux n_f [BOS 13] . The high precision with which n_f can be determined reveals whether the transmembrane water flux is accompanied by solute transport and provides us with some important information about the underlying mechanisms involved in the transmembrane water movements.

5.3. High-content phenotypic screening based on QP-DHM

HCS [BIC 10] is widely used today in academic and pharmaceutical research for investigating a large variety of biological processes through the design phenotypic cellular assays (i.e. based on morphological changes). A series of important screening applications in biological or therapeutic research target the measurement of cell morphological variations [CAR 06] for monitoring events such as cell death mechanisms and cytotoxicity profiling upon interaction with drugs or interfering compounds. For these applications, it is of paramount importance to use the least invasive imaging techniques possible to avoid phenotypic changes induced by the imaging process itself. Therefore, non-invasive and label-free technologies offer a strong advantage in comparison with labeled approaches for which the requirement of exogenous labels may alter the intactness of cells, in addition to the need for extra pipetting or dispensing steps to deliver probes and the required time-consuming image focusing process prior to image acquisition, which are particularly limiting as far as time-lapse experiments are concerned. As illustrated by the different applications presented above, the QPS provides original and unique information about cell morphology and content with a high sensitivity. Consequently, QPM in general and in particular QP-DHM, due to its numerical flexibility, represent an appealing imaging modality to perform high-content phenotypic screening. This view is supported by some studies which have permitted us to stress the possibility to develop *in vitro* label-free cytotoxicity screening assays based on QP-DHM [BAU 11, PAV 12, KUH 13, RAP 14]. Specifically, Pavillon *et al.* [PAV 12] have explored *in vitro* the early detection of excitatory neuronal death induced by perfusion of glutamate, the main excitatory neurotransmitter in the brain (see section 4.2). Practically, the analysis of the early QPS changes, reflecting in particular a rapid neuronal cell volume regulation in response to glutamate, has permitted us to predict, within a time frame of tens of minutes, whether a subsequent neuronal death would occur.

Otherwise, HCS technology requires being both high in content and screening, and thus consists of the parallelization and automation of a test normally conducted in a single Petri dish. For this purpose, 96-, 384- or 1,536-well plates in which cells are cultured (mostly cell lines or primary cell cultures) are typically used. The compounds to test are added to each well of the plate (before or after adding the cells). Practically, one compound is added per well and a few tens to a few million compounds can be tested. Definitively, a few images per well are acquired and automated image analysis characterizes and quantifies the morphological cell changes induced by the specific compound present in each well.

Indeed, although QP-DHM does not have the advantage of signal specificity provided by labeled approaches, it provides a non-invasive highly sensitive QPS allowing us to directly quantify cytotoxicity [KUH 13] and permits a drastic improvement in scanning speed for multi-well plates due to its numerical autofocusing capability. Practically, instead of having to mechanically focus on each well, a refocusing step is performed off line after the acquisition during the numerical reconstruction of the holograms, in parallel to the image analysis and quantification.

Otherwise, standard image processing of QP images allows us to rapidly determine in which regions of the field contain cells and thus to automatically provide relevant information including cell confluence, proliferation or migration. From this, many parameters derived from the QPS can be defined to quantitatively and specifically characterize the phenotypic cell changes mediated by the tested compounds. For illustrative purpose, Figure 5.2 shows a time course of the mean optical thickness (OT= $(\lambda/2\pi)$*QPS) of an HeLa cell population treated with serial dilution of doxorubicin and the corresponding dose-response graph obtained by measuring the area under the curve of the time-lapse date and allowing us to calculate the half maximal effective concentration (EC_{50}).

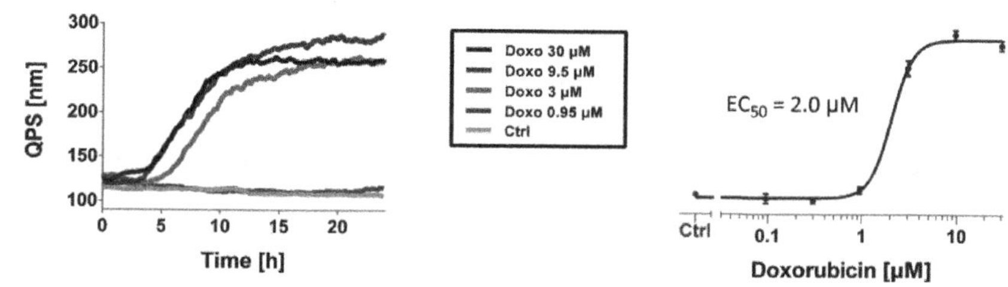

Figure 5.2. *The 24 h time-lapse measurement of the QPS of HeLa cells (a cancer cell line) treated with various concentrations of doxorubicin (an anticancer drug) as well as the corresponding dose response graph allowing us to derive the half maximum effective concentration (EC_{50}). OT = optical thickness defined as the mean $(\lambda/2\pi)$*QPS of HeLa cell population. Modified from [RAP 14]. For a color version of this figure, see www.iste.co.uk/picart/digiholography.zip*

Furthermore, time-lapses yield more information than simple end-point measurements obtained after a given period of time. For instance, we can detect when cells start to respond to the treatment (4–5 h in the case of doxorubicin), when a plateau is reached (here at ~12 h) therefore obtaining a full vision of the dynamics of the process.

In addition to these automated measurements reflecting the behavior of a cell population, a further image analysis aiming to identify relevant quantitative parameters at a single cell level can be easily performed [CAR 06]. This type of analysis allows us, for instance, to distinguish subpopulations of cells showing different phenotypes.

5.4. Multimodal QP-DHM

As illustrated through the different applications presented in the previous sections, QPS being sensitive to both cell morphology and intracellular RI contains a wealth of information, which, however, sometimes requires additional experimental procedures to be exploited in terms of specific and relevant biophysical cell parameters. Furthermore, taking into account that cell processes are multifaceted, obtaining a comprehensible understanding of their underlying mechanisms requires the use of techniques being able to grasp their multiple aspects, ideally in a simultaneous manner. Consequently, the development of QP-DHM or more generally QPM in multimodal approaches represents an important issue in research field of cell biology [PET 10]. Below, three concrete developments are presented combining QP-DHM with epifluorescence, RS and electrophysiology.

5.4.1. *Multimodal fluorescence QP-DHM*

Fluorescence microscopy is a well-established and widely used technique within the field of cell biology and has permitted, in particular, in combination with the utilization of specific fluorescent cellular probes, the observation of dynamic processes in living cells ranging from vesicles fusion to the concentration changes of specific intracellular signaling molecules. Some attempts to combine QP signal with fluorescence imaging have already been published [PAR 06, PAV 12].

Here, we report a multimodal approach combining epifluorescence microscopy with QP-DHM which has permitted us to discuss some important issues concerning cell dynamics by providing simultaneously the measurements of both intracellular ionic homeostasis and cell morphology regulation. Definitively, this multimodal approach has allowed us to study the relationships between the intracellular calcium homeostasis and the cell volume regulation, which plays a critical role in many cell processes

[LAN 98] including cell death processes [PAV 10]. This approach also makes it possible to non-invasively correlate the water fluxes with the dynamics of chosen ion species, which are individually involved in specific membrane receptors and cotransporters. Practically, an imaging module for epifluorescence has been incorporated in a transmission off-axis DHM setup based on a Mach–Zehnder configuration. A dichroic mirror has been introduced in the object arm behind the Microscope Objective (MO) making it possible to excite the specimen in epifluorescence, to reflect the object beam illuminating the specimen onto the charge-coupled device (CCD) camera using to record the holograms and transmit the fluorescence signal emanating from the specimen, which is then detected after wavelength filtering with an electron multiplying CCD camera [PAV 10].

5.4.2. *Multimodal Raman-QP-DHM*

RS is a measurement approach which detects vibrational states of the observed molecules so that it typically probes similar energy levels as infrared spectroscopy, but retains a higher spatial resolution through its excitation in the visible range by relying on the inelastic scattering of light. When used in conjunction with laser-scanning microscopy, RS can then provide hyperspectral images, where each pixel contains the spectral response at the location of the tightly focused excitation spot. In the context of cell imaging, the analysis of Raman spectra can, for example, lead to the classification of undifferentiated cells [GHI 12], or the monitoring of molecular dynamics [OKA 12], among various applications. RS is, however, limited by the small cross section of biomolecules, hence requiring long exposure times (typically, in the seconds range) to be detectable. Stronger signals leading to high-quality images can also be obtained through other approaches, based on nonlinear effects such as coherent anti-Stokes Raman spectroscopy (CARS) or stimulated Raman scattering (SRS) [CHU 13], but these techniques usually provide smaller spectral ranges, with a reduced spectral resolution compared to linear Raman detection, making spectroscopic analysis more challenging.

Compared to other techniques, Raman microspectroscopy is typically difficult to implement in a multimodal system due to its weak unpolarized response, which lies across a wide spectral range (approximately 100 nm), which makes channel separation complicated [PAV 14]. In this context, the low-power interferometric measurement approach of DHM makes it an ideal candidate for combination with RS.

The multimodal system can be implemented through a spectral separation scheme, where the DHM laser wavelength is chosen outside the Raman emission range [PAV 13], so that both signals can be separated through standard dichroic optical components. This acquisition approach makes it possible to retrieve simultaneously both signals, which provide very different features, while retaining the label-free capability of the measurement. On the one hand, DHM enables real-time acquisition, which can be used to observe rapid morphological changes of cell bodies during Raman recording, which typically requires several minutes for a hyperspectral image. On the other hand, RS retrieves a highly specific measurement, providing insights about the intracellular molecular content (lipids, proteins, DNA, ribonucleic acid (RNA), etc.) that phase shifts measurements cannot unravel.

An illustration of simultaneous acquisition is presented in Figure 5.3, where live unstained macrophage-like Raw264.7 cells are observed in both channels. The QP images (Figure 5.3(a)) provide quantitative information based on the intracellular RI in real-time, making it possible to monitor the morphological changes during the Raman acquisition (here at $t = 0$, 7 and 15 min). It is typically possible to identify the movement of cells at the center of the field of view, or the development of small protrusions, which are not visible in the slower Raman channel. On the other hand, the Raman hyperspectral stack can be used to extract various image contrasts (Figure 5.3(b)) by highlighting the right bands. It is, for example, possible to generally highlight biomolecules with a C-H stretching band (Figure 5.3(b1) green, 2,940 cm^{-1}) or lipids (Figure 5.3(b2) red, 2,870 cm^{-1}). We can also extract information about more specific molecules, such as proteins (Figure 5.3(b2) green, 1,680 cm^{-1}) or cytochrome c (Figure 5.3(b1) red, 750 cm^{-1}), which provides a strong signal as it is resonant at the excitation wavelength (532 nm). Typical spectra for various cell locations are shown in Figure 5.3(c), showing some spectral differences which lead to image contrast.

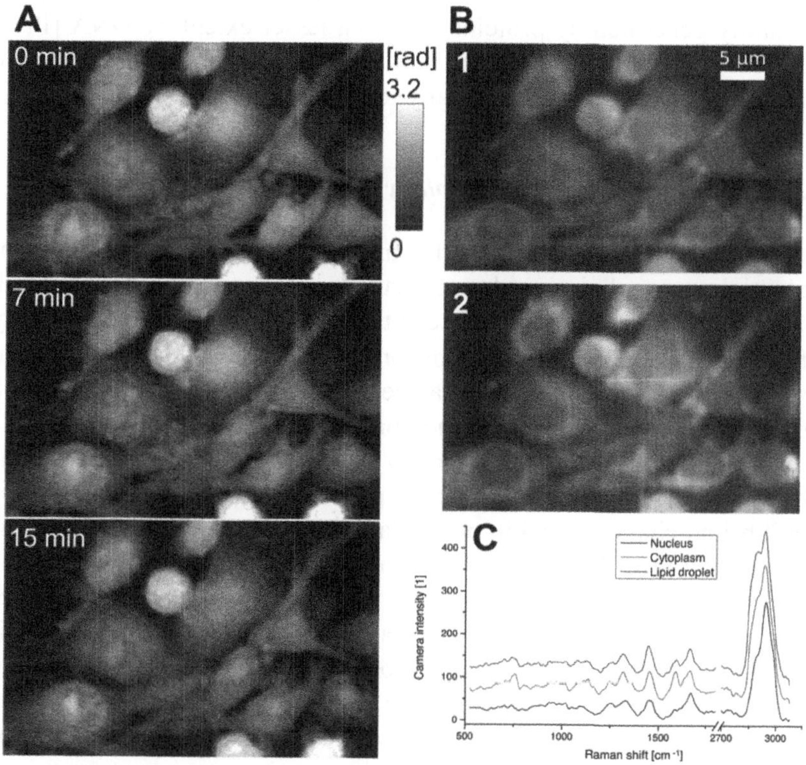

Figure 5.3. *Macrophage-like Raw264.7 cells observed both a) in DH-QPM with a faster acquisition rate (here shown at 0, 7 and 15 min) and b) in Raman microscopy with various molecular contrasts highlighted ((b1) green: C-H stretching (2,940 cm^{-1}), red: cytochrome c (750 cm^{-1}); (b2): green: proteins (1,680 cm^{-1}), red: lipids (2,840 cm^{-1})). c) Typical Raman spectra from various locations in cells. For a color version of this figure, see www.iste.co.uk/picart/digiholography.zip*

As both QP and Raman images are label-free methods probing intracellular molecules, it is interesting to identify the differences between the images. However, they both rely on different physical interactions: while DHM measures the complex wave field linearly scattered by the specimen, the Raman effect is based on the inelastic scattering of the excitation light. By then analyzing the spectral information through multivariate analysis, it is possible to break down the different molecular contributions to the QP signal, showing that it is mainly generated by proteins, and to a lesser extent by DNA [PAV 13]. On the other hand, lipids, which have a very strong signal in Raman, contribute more marginally to phase.

5.4.3. *Multimodal electrophysiology QP-DHM*

Cells in general and neurons in particular present a small voltage or potential across their membranes called resting membrane potential. Basically, the resting potential results from the selective permeability of cell membrane to the different ions including Na^+, K^+ and Cl^- and from the presence of intracellular negatively charged proteins, which are denied passage across cell membrane. Practically, this ion-specific membrane permeability is due to the presence of specific ions channels. Otherwise, the resting potential represents a dynamic equilibrium characterized in particular by continuous ion fluxes across the membrane. Definitively, each ion species, for which the equilibrium potential resulting from its electrochemical gradient does not correspond to the resting membrane potential, contributes to this ion flux. In addition, the concentration gradient of the different type of ions is maintained by the presence in the membrane of active ion transports including pumps which use metabolic energy to pump ions across their concentration gradient.

Practically, electrophysiological approaches, and in particular patch-clamp techniques, represent the gold standard for studying membrane potentials and the transmembrane currents. A comprehensive description of such approaches is far beyond the scope of this chapter but can be found in many reference books. However, in order to facilitate reading and comprehension of the text, we summarize here some basic concepts of patch-clamp recording, and in particular the whole cell mode which is the most commonly used. Concretely, in this configuration, an open tip thin glass micropipette, containing both an ionic solution mimicking the intracellular solution and a chlorided silver electrode connected to a highly sensitive electronic amplifier, is pressed against a cell membrane. A brief, strong suction is applied to both assist in the formation of a high-resistance seal between the glass and the cell membrane and to disrupt the membrane patch. The interior of the pipette becomes continuous with the cytoplasm and a bath electrode is used to set the zero level. In whole-cell measurements, the two main configurations are: 1) voltage-clamp mode and 2) current-clamp mode. In the voltage-clamp mode, the cell membrane potential is kept constant via a negative feedback loop injecting a compensating current into the cell that resembles the current that is flowing through the whole-cell membrane. Practically, the recording of this compensating current provides information about the whole-cell membrane conductance. In addition, the current–voltage relationships can be investigated by considering that the membrane potential can be manipulated independently of ionic currents. In contrast, in the current-clamp mode, the injected current is kept constant and changes in the membrane potential can be measured.

As seen in section 2.6, even at constant extracellular osmolarity, the volume constancy of any mammalian cell is permanently challenged by its normal activity, including in particular the electrical activity. Consequently, combining electrophysiology and QP-DHM, which have, in particular, the capacity to monitor cell volume as well as transmembrane water movements, is likely to provide a more comprehensive understanding of the mechanisms associated with the electrical activity, which challenge cell volume.

Practically, we have developed a multimodal system combining electrophysiology in a whole-cell configuration and QP-DHM aiming at specifically measuring the transmembrane water fluxes associated with cell electrical activity. Experiments using a well-established biological model involving Cl^- conductance – a human embryonic kidney (HEK) cell line transfected to express the ligand-gated chloride ion channel called gamma-aminobutyric acid ($GABA_A$), the acronym of its endogenous ligand, the gamma-aminobutyric acid – have been performed. Due to these experiments, it has been established that a net ionic influx corresponds to a phase decrease due to the osmotic water entry, while a net ionic efflux induces a phase increase reflecting the osmotic exit of water. Furthermore, under the following two assumptions – 1) the n_c value depends linearly on the concentration of the different intracellular components according to the seminal work of Davies and Wilkins [DAV 52] and 2) the volume variations induced by the transmembrane water movements accompanying the ionic movements are proportional to the number of ions having passed through the membrane or channels – it has been possible to establish the following relation between QPS $\varphi(t)$ and transmembrane current I(t) [JOU 12]:

$$I(t) = \frac{V_0}{\varepsilon}\frac{d}{dt}\left(\frac{\varphi_0}{\varphi(t)}\right)^{1/s-r} - \frac{1}{\varepsilon}\frac{d\Delta V(t)}{dt} \qquad [5.3]$$

with V_0 being the cell volume measured at the beginning of the experiments at a time $t = t_0$, ΔV the non-electrogenic volume changes corresponding to any cell volume variation except those directly related to a transmembrane net current, ε a constant representing the volume variation associated with the net charge movement through the cell membrane [ml/C] and r a parameter related to the cell deformation associated with the volume variation and the parameter s which takes into account cell processes inducing volume variations or not but leading to intracellular RI changes by mechanisms other than dilution or concentration of the intracellular content by ions and water fluxes. Practically, $r \in [0,1]$ with $r = 1/3$ and $r = 1$ corresponding to isotropic volume changes and piston-like cell movements in the z direction, i.e. along the optical axis, respectively. Usually, the adhesion of cultured cells on a coverslip facilitates a movement along the z-axis resulting in $r \in [0.5, 0.8]$, typically. In contrast, the case $s = 1$

corresponds to a simple dilution or concentration of the cell content, while $s < 1$ reflects modification of intracellular RI by processes mediating release or cellular internalization of substrates including protein release mechanisms or cotransport activation involved in the volume regulatory processes. The parameters V_0, r and s can be determined by the decoupling procedure described in section 2.6. Consequently, two parameters remain to be determined, namely ε and ΔV, to calculate the current.

Experiments with HEK cells have validated equation [5.3] regarding Cl− currents mediated by GABA applications (Figure 5.4(a_1)) with ε_{GABA} lies within the range of 90–110 $\mu m^3/nC$, and have permitted to construct a phase response as a function of the clamped membrane potential (φ / V curve) allowing us to optically determine the equilibrium potential for Cl^- with the same precision as an electrophysiological approach measuring the classical I/V curves (Figure 5.4(a_2)) [JOU 12]. Practically, ε_{GABA} has been estimated for a membrane potential clamped at –40 mV by performing a least-squares fit to adjust the phase current (Pred. Curr.) derived from equation [5.3] with the measured current (electrophysiological recordings in a voltage-clamp mode). Using this ε_{GABA} determination, Pred. Curr. has been calculated for other values of the membrane potential by using equation [5.3] (Figure 5.4(a_1)).

This relationship between QPS and transmembrane currents affords, through appropriate mathematical treatments of the optical signal, the possibilities to non-invasively resolve the neuronal network activity specifically (see section 5.5) and the dynamics of the current triggered by the ionotropic receptor activity generally. This approach is thus amenable to use for pharmacological screenings of modulators developed for the management of human pathologies involving dysfunctions of specific inotropic channel including chloride channels associated with cystic fibrosis [JOU 14]. Furthermore, the important issue concerning the quantitative monitoring of the transmembrane water fluxes associated with the activity of various electrogenic or non-electrogenic cotransport can also be relevantly discussed [JOU 11].

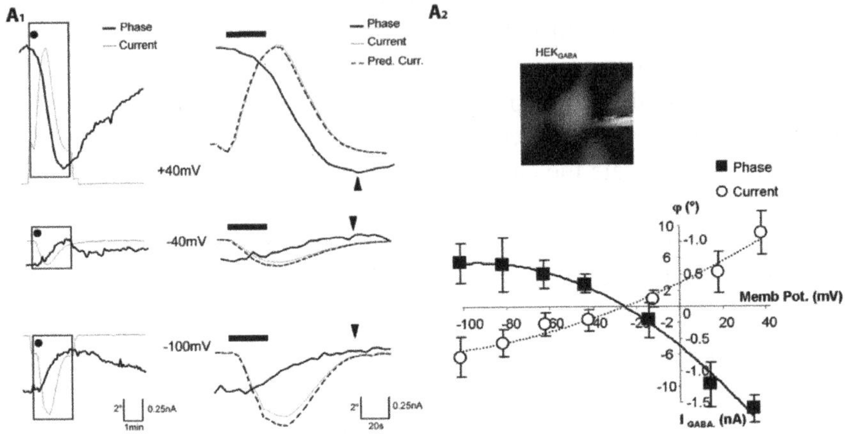

Figure 5.4. *a) HEK cells and GABA applications and b) neurons in response to glutamate. (a_1) (left) Example of three simultaneous traces of current (gray line) and phase shift (black line) recorded with 44 mM of $[Cl^-]_{intrapip}$ on the same HEK_{GABA} after application of GABA (3 μM; 30 s; dot). At –100 mV, application of GABA triggered an inward current concomitantly to an increase in the phase signal. Conversely, at +40 mV, same applications of GABA triggered an outward current accompanied by a decrease in the phase signal. Note that for –40 mV (close to the resting potential for Cl), the current and the phase shift were very small. (Right) Expansion of traces visualized in A1 (parts defined by rectangles). For each level of membrane potential, the peak of phase shift (indicated by the arrow head) was reached when the I_{GABA} was terminated. According to equation [5.3], the phase signal can predict the current (Pred. Curr.: dashed line) superimposed to the recorded current. (a_2) The φ/V curve (full square and thick line) and the I/V curve (empty circle and thin line) obtained with GABA (n = 7) indicated an E_{Cl} of –26 and –26 mV, respectively. (Inset) Representative quantitative phase image of patched an HEK cell*

5.5. Resolving neuronal network activity and visualizing spine dynamics

5.5.1. *Background*

Within the brain, water is distributed between blood, cerebrospinal fluid and interstitial and intracellular compartments, and moves across these compartments following the differences in osmotic and hydrostatic pressures. At constant hydrostatic pressure gradients, water movements depend largely on the osmotic gradients, which are created by the concentration of osmolytes, including ions, amino acids and derivatives such as neurotransmitters, at the extracellular and intracellular compartments. At the microscopic level, water transport is involved in cell volume regulation and in controlling the dimensions of the extracellular space (ECS) [SYK 03, NAG 04]. However, neuronal activity challenges the homeostasis of these intra- and extracellular compartments. Indeed, considering a low intracellular concentration of Na^+, resulting from both a neuronal membrane with a low permeability to Na^+ and sodium–potassium transporters continuously pumping out Na^+, if suddenly, the Na^+ conductances increase, it will cause Na^+ to rush into a neuronal cell. This movement of Na^+ ions along their concentration gradient is also enhanced by the transmembrane electrostatic gradient corresponding to a negative resting membrane potential (–70 mV) due mainly to a large quantity of negatively charged intracellular proteins. This inward flow of Na^+ ions changes the electrochemical gradient, which, in turn, produces a further rise in the membrane potential called a depolarization. This mechanism is precisely what initiates the action potential. Otherwise, the Na^+ currents flowing inward at a point on the axon during an action potential spread out along the axon and depolarize the adjacent sections of its membrane. When actions potentials are conducted down an axon – their size remains undiminished as they travel – they finally induce a depolarization in a presynaptic terminal button, the end of an axonal branch, which generally faces the postsynaptic membrane of the neuron that receives the message. The two membranes face each other across the synaptic cleft with a gap approximately 20 nm wide. Practically, when the depolarization reaches the terminal button, small vesicles, containing molecules of the neurotransmitter and located just inside the presynaptic membrane, fuse with the membrane and then break down, spilling their contents into the synaptic cleft. The neurotransmitter molecules thus released from the terminal button, then diffuse across the synaptic cleft. When they reach and attach postsynaptic membrane, neurotransmitter–receptor ion channels begin to open, which permits the passage of specific ions into or out the cell. Consequently, the local membrane potential of the postsynaptic neuron is modified, which could, in turn, induce or inhibit the generation of action potentials [CAR 14].

Consequently, these transmembrane ions movements and neurotransmitter release including glutamate, glycine and GABA related to the neuronal activity are accompanied by transmembrane water movements for osmotic reasons. The result is that the regulation of both the ECS and the cell volume is affected by the level of neural activity.

On the other hand, a distinct feature of nervous tissue is the intricate network of synaptic connections among neurons with diverse morphologies. Although initial connections are formed largely through molecular mechanisms, it has been clearly established that electrical activities influence not only the structure and efficacy of existing synapses but can also contribute to the production of new synapses, in particular through changes in the size and shape of dendritic spines and the formation of new ones. Indeed, synapses can occur mainly in three places of the postsynpatic neuron: on the soma, on others axons and on the smooth surface of dendrites – branched projections which compose most of the receptive surface of a neuron –, or on dendritic spines – small membranous protrusions, with typical sizes of a few hundred nanometers, which stud the dendrites. The axodendritic synapses represent the largest proportion of the excitatory synapses in the mammalian brain and the glutamate is the most common excitatory neurotransmitter, inducing a depolarization of the postsynaptic neurons particularly by opening the *N*-methyl-d-aspartate (NMDA), the 2-amino-3-(3-hydroxy-5-methylisoxazol-4-yl) proprionate (AMPA) and the kainate ionotropic receptors, representing the three main classes of glutamate-receptor ion channels. All these abilities of synapses to strengthen, weaken or to be created over time, in response to increases or decreases in the neuronal network activity, called synaptic plasticity, have been intensively studied in recent years significantly stimulated by the development of new optical methods, which have in particular permitted to resolve the morphology and the dynamics of dendritic spines [PAD 11].

Furthermore, plasticity mechanisms affect information transfer on multiple time scales and levels of specificity and play a pivotal role in integrative function and learning. However, with a few exceptions, including activity-induced long-term potentiation (LTP) and long-term depression (LTD) and some specific activity-dependent mechanisms that can modulate the density of dendritic spines onto which the majority of fast excitatory synaptic transmission operate, our understanding of how electrical activity affects the structure and function of neuronal networks is very limited. Within this framework, the development of techniques that provide a non-invasive resolution of both local neuronal network activity and dendritic spine dynamics is highly relevant. In light of this, we present how QP-DHM, due to its possibilities of both 3D high resolution through tomographic developments and monitoring cell volume and transmembrane water fluxes, can start in combination with epifluorescence microscopy and/or electrophysiology to discuss these two issues in an innovative way.

5.5.2. *Imaging neuronal activity by measuring transmembrane water movements with QP-DHM*

With respect to the study of neuronal activity, electrophysiological approaches, in particular patch-clamp techniques, have permitted major breakthroughs leading to the discovery of LTP and LTD, by setting a voltage across the neuronal membrane and directly measuring the current flowing through a single ion channel. Basically, patch clamp is the gold standard for assessing ion channel function allowing discriminating ionic currents in the femtoampere (10–15 fA) range and with microsecond time resolution (see section 4.3). However, patch clamp is still a highly invasive and laborious process requiring precise micromanipulations and a high degree of operations, which generally imposes the recording of voltages on a limited number of cells that form a neuronal network. Optical techniques, however, seem to be an ideal solution for measuring membrane potentials since they are relatively non-invasive and could work both at low and high magnification. For instance, calcium indicators used in combination with high resolution, two-photon microscopy allows performing measurements of the spiking activity from hundreds to several thousand neurons in mammalian circuits while still keeping track of the activity of each neuron individually [COS 03, PET 11]. However, calcium imaging has its shortcomings and cannot substitute for voltage imaging [PET 11]. Practically, voltage imaging methods have lagged behind calcium imaging due to important challenges related to physical constraints of the measurements themselves, including an electrical field highly located to the thin membrane region, which is essentially a two-dimensional plasma membrane which cannot contain an arbitrary number of voltage chromophores without disrupting its properties, as well as the fact that the plasma membrane represents only a small proportion of the total membrane surface in the neuron on which chromophores are attached. Finally, the relatively high speed of the electrical responses of mammalian neurons also represents a serious challenge for voltage measurements. Consequently, despite some promising perspectives, the different voltage imaging methods suffer from poor signal-to-noise and secondary side effects, and have up to now fallen short of providing single-cell resolution when imaging the activity of neuronal populations [PET 11].

Otherwise, it is well known that the neuronal activity induces modifications of the intrinsic optical properties at the subcellular [HIL 49, TAS 68, COH 73, CAR 04], cellular [STE 91, TAS 92] and tissue level [MAC 91, AND 96, HOL 96]. Consequently, we have attempted to study the neuronal activity with QP-DHM due to its capacity to monitor transmembrane water movements as described in sections 2.6 and 4.3. Basically, equation [5.3] established in the context of Cl^- current measurements (see section 4.3) has been used to optically monitor neuronal network activity mediated by glutamate application, the main excitatory neurotransmitter in the brain released at 80% of synapses. Concretely, the early stage of neuronal responses induced by glutamate on primary culture of mouse cortical neurons has been explored with an approach combining QP-DHM and electrophysiology. This study involving pharmacological experiments revealed that glutamate produces three distinct optical responses, predominantly mediated by NMDA receptors: biphasic (Biph.), reversible decrease (RD) and irreversible decrease (ID) responses (Figure 5.5(a_3)).

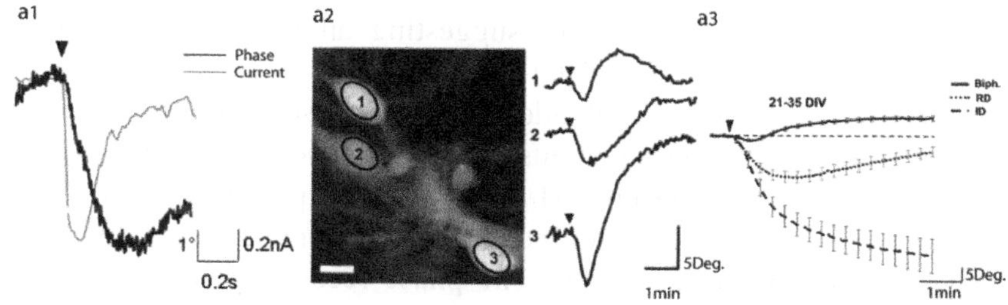

Figure 5.5. *a) Local application of a short glutamate pulse (500 μM, 200 msec; arrow head) on a patched neuron triggered both a strong transient decrease in the phase signal associated with an inward current. Phase is expressed in degrees. b) Application of glutamate triggers three main types of phase shifts: (left) representative quantitative phase image of three cultured cortical neurons visualized in DHM. Ovals in the middle of the cells corresponded to the region of interest for the phase signal (scale bar, 10 μm). (Right) application of glutamate (30 μM, 30 sec; arrowhead) triggered different types of shapes of optical signals from these three recorded neurons. c) Averaged traces of the three characteristic optical signals induced by glutamate application (30 μM, 30 sec, arrowhead) and recorded from 21–35 DIV neurons. These phase signals are classified into three categories: a biphasic response (Biph.; n = 137), a reversible decrease in the phase signal (RD, n = 80) and an irreversible decrease in the phase signal (ID, n = 35). All these experiments were performed at room temperature*

The shape and amplitude of the optical signal were not associated with a particular cellular phenotype but reflected the physiopathological status of neurons linked to the degree of NMDA activity. Thus, the “Biph.”, “RD” and “ID” responses indicated, respectively, a low-, high- and “excito-toxic” level of NMDA activation. For these three distinct responses, a strong inward current associated with the early phase decrease, whose amplitude is proportional to the concentration of glutamate and the duration of the application, has been measured. This inward current is consistent with glutamate-mediated activation of specific ionotropic receptors including the NMDA, AMPA and the kainate ionotropic receptors, whose opening induces influxes of Ca^{2+} and Na^{+} down their electrochemical gradient. Consistently, the phase decrease results from water entrance accompanying influxes of Ca^{2+} and Na^{+} for osmotic reasons. Moreover, furosemide and bumetanide, two inhibitors of sodium- and/or potassium-coupled chloride movement, strongly modify the phase shift suggesting an involvement of two neuronal cotransporters, Na-K-Cl (NKCC1) and K-Cl (KCC2), in the genesis of optical signal. In addition, the phase responses can be decomposed into two components: a rapid component accompanying glutamate-mediated current (I_{GLUT}) – the phase decrease in Figure 5.5(a) and a slow component corresponding generally to a phase recovery while $I_{GLUT} = 0$. The phase recovery, which is much slower than the fast one, is likely to correspond to a non-electrogenic neuronal volume regulation involving several mechanisms. Interestingly, the time course presented in Figure 5.5(a) shows that the water movements are not significantly delayed – at the tenth of second scale at least – relative to the recorded current. The measurements of I_{GLUT} as well as the corresponding intracellular RI and neuronal volume changes allowed us to estimate the parameter "$\varepsilon_{\text{GLUT}}$" [mL/C]. Practically, values of ε_{GLUT} lie within the range of 60–120 μm^3/nC equivalent to 340–620 water molecules transported per ion having crossed the membrane. Interestingly, the typical intracellular RI change induced by a glutamate pulse (500 µM, 0.2 s) is approximately 0.002–0.003. The associated neuronal swelling is approximately 100 femtoliter (fL) for a typical neuronal cell body of 1,500 fL corresponding to a 6–7% cell volume variation. However, these orders of magnitude correspond to an exogenous glutamate application during a few tenths of a second. Physiological release of endogenous glutamate is likely to induce smaller water movements and intracellular RI changes. Finally, it is important to note, although we do not yet have a comprehensive understanding of the underlying mechanisms inducing these phase responses, that these consequent transmembrane movements of water accompanying the neuronal activity are in good agreement with the typical amount of water fluxes induced by various cotransporters including NKCC1 [HAM 05].

Furthermore, information concerning transmembrane currents obtained from electrophysiological recordings combined with QP-DHM signal allows us to pave the way for developing a simultaneous multiple sites optical recording of transmembrane currents capable to resolve local neuronal network activity [JOU 12].

5.5.3. *3D Visualization of dendritic spine dynamics with quantitative phase tomographic microscopy (QP-TM)*

As previously mentioned, dendritic spines are the postsynaptic components of most excitatory synapses in the mammalian brain and are the loci of structural and functional plasticity. In past decades, studies have revealed that the number and size of dendritic spines are regulated by a variety of gene products and environmental factors, underscoring the dynamic nature of spines and their importance to brain plasticity. Because dendritic spines are the key elements for information acquisition and retention, understanding how spines are formed and maintained, in relation with neuronal network activity, will likely provide fundamental insights into how the brain possesses the extraordinary capacity to learn and remember [BHA 09]. However, dendritic spines are tiny transparent processes whose visualization and resolution remain a challenge for optical approaches. Nevertheless, the advent of two-photon (2P) excitation laser scanning microscopy, despite limitations inherent to the use of fluorescence, has remarkably contributed to the understanding of dendritic spine function and plasticity [SVO 06, LEE 12].

On the other hand, the diffraction tomography presented in Chapter 1 is particularly efficient when the wavelength of the irradiating beam is comparable to the dimensions of the diffracting parts of the object. It thus represents a promising approach to dynamically and non-invasively resolve subcellular structures including dendritic spines. Indeed, appealing applications to cell biology have already been obtained and published. Practically, the investigation of the RI 3D map has been published for pollen grains [CHA 06a], amoeba [CHA 06b], human RBCs parasitized by plasmodium falciparum [PAR 10b], human influenza virus infected cells and epithelial cheek cells and granulocytes [SIM 10]. Images of HeLa cells at 0° and up to 50° illumination incidences have been published [KIM 12] and DIC have been derived from synthetic aperture PhC microscopy. Interestingly, an approach based on the rotation of the sample using a hollow optical fiber as a live-cell imaging chamber has permitted us to measure the RI 3D maps of cell clusters [KUS 14]. Recently, Cotte *el al.* [COT 13] could firmly establish the benefit of synthetic aperture elevated to quasi-2π by pushing the occupation of the Fourier space up to a diameter equal to the diameter of the Ewald sphere. Superresolution could be obtained and details as small as 100 nm can be observed. Moreover, dynamic images showing the detailed evolution of dendrite spines (Figure 5.6) have been obtained when glutamate is applied.

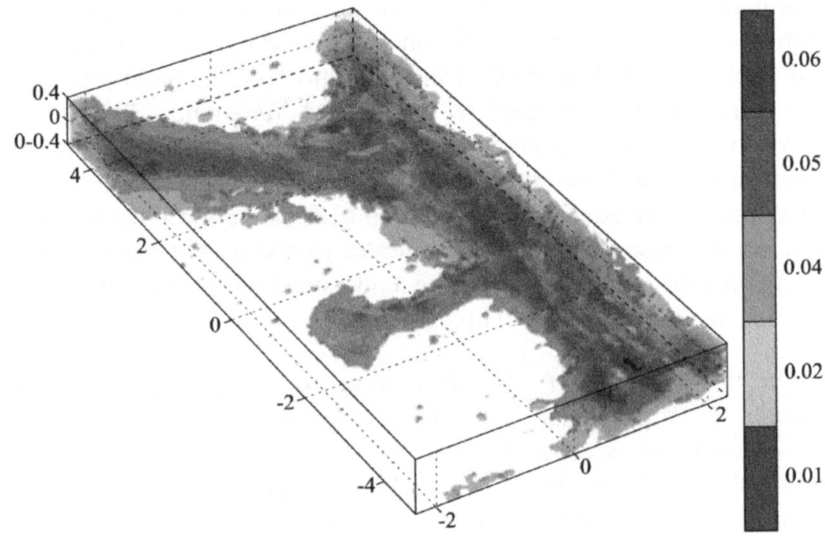

Figure 5.6. *3D representation of a tomographic image of a neuronal process obtained by synthetic aperture quantitative phase image. The details of protruding dendritic spines can be visualized. Scale is in micrometer and the color code represents the difference of refractive index (from 0.04 (red) to 0.16 (cyan)] compared with the extracellular medium (Courtesy of Yann Cotte). These experiments were performed at room temperature. For a color version of this figure, see www.iste.co.uk/picart/digiholography.zip*

5.6. Perspectives

As illustrated through these different applications, QPM and specifically DH-QPM due to the unique possibilities offered by the numerical reconstruction and propagation of the object wave front including extended depth of focus and *a posteriori* numerical refocusing allows us to quantitatively and non-invasively study cell structure, dynamics as well as content at different time scales ranging from a few milliseconds to several days. However, although the QPS is highly sensitive and contains a wealth of information, its interpretation in terms of specific biological processes remains an issue. Practically, discussing this interpretation issue beyond the well-established calculation of the cell DM involves the ability to calculate from the phase signal relevant biophysical cell parameters including absolute volume, membrane fluctuations at the nanoscale, biomechanical properties, transmembrane water permeability and current. Definitively, the derivation of these cell biophysical parameters often requires to be able to obtain separately information relative to both cell morphology (thickness) and content (intracellular RI). Practically, as we have seen, this requires the development of experimental procedures which are sometimes not very convenient to implement. Within this framework, the development of DHM-based optical diffraction tomography, providing direct access to the 3D map of the intracellular RI, could allow us to revisit this difficulty of interpretation. Future developments leading to the real-time and high-resolution 3D maps of the intracellular RI could provide, in addition to the possibility of non-invasively resolving both neuronal network activity and dendritic spine dynamics as presented in section 5.5, invaluable information about cytoarchitecture and compartmentalization of cytoplasm, which plays a critical role in several fundamental cell mechanisms, including proteins synthesis.

Furthermore, the developments of QP-DHM and QP-TM in multimodality approaches, opening up the possibility to simultaneously measure a large number of relevant and specific cell parameters, are particularly promising when considering the multifaceted characteristics of physiology and pathophysiology cellular processes. In addition to the presented multimodal approaches, multimodality combining QP-DHM with coherent anti-Stokes RS (CARS), second harmonic generation (SHG) and third harmonic generation (THG) are particularly promising to provide a large number of relevant and specific parameters yielding an extended cell profiling approach. Indeed, SHG [CAM 11] and THG [FUJ 08] convey highly relevant information concerning in particular membrane proteins and intracellular constituents. Due to the highly nonlinear character of these light generating processes, local properties of cytoskeleton and membrane are obtainable, as well as 3D cellular arrangements by taking advantage of the remarkable penetration power of the SHG or, more particularly, the THG processes. Otherwise, used in conjunction with Atomic force microscopy (AFM), QP-DHM would provide an image of the local deformations and therefore a more detailed and informative insight of biophysical properties of the cell. In addition, the use of free wandering particles would also inform precisely on the nature of binding forces and properties of the macromolecules.

This detection of multiple biophysical cell parameters resulting from multimodal approaches could even be scaled into a high-throughput cellular screening assay [KUH 13], which allows the monitoring of several cellular processes concurrently as well as their modulations as a function of the transcriptome and metabolome of different cell types. This paves the way to explore cell biomarkers of risk for different diseases including psychiatry disorders. Indeed, recent years have seen a renewed interest for cell biomarkers due to the possibility, through induced pluripotent stem cells, to convert somatic cells (e.g. fibroblast) obtained from patients, controls as well as high-risk subjects (skin biopsy) into neurons (grown in culture) with the genetic makeup necessary for the development of certain phenotypes of the disease. Although this approach has currently several limitations [WU 11], it allows us to consider the hypothesis that some phenotypes are already found *in vitro*, at the cell level or in a network of cells taken from patients, a “cell phenotype”, is particularly attractive [BRE 12].

5.7. Acknowledgments

The work presented in this review is the result of a close collaboration between the Microvision and microdiagnosis group (SCI/STI/CHD group) of EPFL with Florian Charrière, Jonas Kühn, Etienne Shaffer, Fatih Toy, Yann Cotte and Christian Depeursinge, the Laboratory of Neuroenergetics and Cellular Dynamics, Brain and Mind Institute of EPFL with Pierre Magistretti, Benjamin Rappaz, Pascal Jourdain, the Center for Psychiatric neuroscience, CHUV with Daniel Boss, Kaspar Rothenfusser and the company Lyncée Tec SA, Lausanne (http://www.lynceetec.com). We thank the Swiss National Science Foundation (SNSF) grants no. CR3213_132993 the National Center of Competence in Research Synapsy (http://www.nccr-synapsy.ch/) as well as the foundation de Préfargier.

5.8. Bibliography

[ANA 11] ANAND A., CHHANIWAL V.K., JAVIDI B., "Imaging embryonic stem cell dynamics using quantitative 3-D digital holographic microscopy", *IEEE Photonics Journal*, vol. 3, no. 3, pp. 546–554, 2011.

[AND 96] ANDREW R.D., ADAMS J.R., POLISCHUK T.M., "Imaging NMDA- and kainate-induced intrinsic optical signals from the hippocampal slice", *Journal of Neurophysiology,* vol. 76, no. 4, pp. 2707–2717, 1996.

[ANS 12] ANSARI R., AHERRAHROU R., AHERRAHROU Z., *et al.*, "Quantitative analysis of cardiomyocyte dynamics with optical coherence phase microscopy", *Optical Coherence Tomography and Coherence Domain Optical Methods in Biomedicine XVI, Proc. of SPIE 8213*, vol. 821338-5, 2012.

[ANT 08a] ANTKOWIAK M., CALLENS N., SCHOCKAERT C., *et al.*, "Accurate three-dimensional detection of micro-particles by means of digital holographic microscopy – art. no. 699514", in GORECKI C., ASUNDI A.K., OSTEN W. (eds.), *Optical Micro- and Nanometrology in Microsystems Technology II, Proceedings of SPIE 0277-786X*, vol. 6995, pp. 99514–99514, 2008.

[ANT 08b] ANTKOWIAK M., CALLENS N., YOURASSOWSKY C., *et al.*, "Extended focused imaging of a microparticle field with digital holographic microscopy", *Optics Letters,* vol. 33, no. 14, pp. 1626–1628, 2008.

[BAE 10] BAE Y., LEE S., YANG W., *et al.*, "Three-dimensional single particle tracking using off-axis digital holographic microscopy", *Nanoscale Imaging, Sensing, and Actuation for Biomedical Applications II, Proc. SPIE 7574*, vol. 7574, p. 757408, 12 February 2010.

[BAR 52] BARER R., "Interference microscopy and mass determination", *Nature*, vol. 169, no. 4296, pp. 366–367, 1952.

[BAU 11] BAUWENS A., BIELASZEWSKA M., KEMPER B., *et al.*, "Differential cytotoxic actions of Shiga toxin 1 and Shiga toxin 2 on microvascular and macrovascular endothelial cells", *Thromb Haemost*, vol. 105, no. 3, pp. 515–528, 2011.

[BER 08] BERNHARDT I., IVANOVA L., LANGEHANENBERG P., *et al.*, "Application of digital holographic microscopy to investigate the sedimentation of intact red blood cells and their interaction with artificial surfaces", *Bioelectrochemistry*, vol. 73, no. 2, pp. 92–96, 2008.

[BER 09] BERGOEND I., COLOMB T., PAVILLON N., *et al.*, "Depth-of-field extension and 3D reconstruction in digital holographic microscopy", in BOSSE H., BODERMANN B., SILVER R.M. (eds.), *Modeling Aspects in Optical Metrology II, Proceedings of SPIE 0277-786X*, vol. 7390, 2009.

[BHA 09] BHATT D. H., ZHANG S., GAN W.B., "Dendritic spine dynamics", *Annual Review of Physiology*, vol. 71, pp. 261–282, 2009.

[BIA 13] BIANCO V., PATURZO M., GENNARI O., *et al.*, "Imaging through scattering microfluidic channels by digital holography for information recovery in lab on chip", *Optics Express*, vol. 21, no. 20, pp. 23985–23996, 2013.

[BIC 10] BICKLE M., "The beautiful cell: high-content screening in drug discovery", *Analytical and Bioanalytical Chemistry*, vol. 398, no. 1, pp. 219–226, 2010.

[BOH 11] BOHM M., MASTROFRANCESCO A., WEISS N., *et al.*, "PACE4, a member of the prohormone convertase family, mediates increased proliferation, migration and invasiveness of melanoma cells *in vitro* and enhanced subcutaneous tumor growth *in vivo*", *Journal of Investigative Dermatology*, vol. 131, pp. S108–S108, 2011.

[BOS 12] BOSS D., HOFFMANN A., RAPPAZ B., *et al.*, "Spatially-resolved eigenmode decomposition of red blood cells membrane fluctuations questions the role of ATP in flickering", *Plos One*, vol. 7, no. 8, p. e40667, 2012.

[BOS 13] BOSS D., KUHN J., JOURDAIN P., *et al.*, "Measurement of absolute cell volume, osmotic membrane water permeability, and refractive index of transmembrane water and solute flux by digital holographic microscopy", *Journal of Biomedical Optics*, vol. 18, no. 3, pp. 036007, 2013.

[BRE 12] BRENNAND K.J., SIMONE A., TRAN N., *et al.*, "Modeling psychiatric disorders at the cellular and network levels", *Molecular Psychiatry*, vol. 17, no. 12, pp. 1239–1253, 2012.

[BYU 12] BYUN H., HILLMAN T.R., HIGGINS J.M., *et al.*, "Optical measurement of biomechanical properties of individual erythrocytes from a sickle cell patient", *Acta Biomaterialia*, vol. 8, no. 11, pp. 4130–4138, 2012.

[CAM 11] CAMPAGNOLA P., "Second harmonic generation imaging microscopy: applications to diseases diagnostics", *Analytical Chemistry*, vol. 83, no. 9, pp. 3224–3231, 2011.

[CAR 04] CARTER K.M., GEORGE J.S., RECTOR D.M., "Simultaneous birefringence and scattered light measurements reveal anatomical features in isolated crustacean nerve", *Journal of Neuroscience Methods*, vol. 135, nos. 1–2, pp. 9–16, 2004.

[CAR 06] CARPENTER A.E., JONES T.R., LAMPRECHT M.R., *et al.* "CellProfiler: image analysis software for identifying and quantifying cell phenotypes", *Genome Biology*, vol. 7, no. 10, p. R100, 31 October 2006.

[CAR 11a] CARDENAS N., YU L.F., MOHANTY S.K., "Probing orientation and rotation of red blood cells in optical tweezers by Digital Holographic microscopy", *Optical Diagnostics and Sensing XI: Toward Point-of-Care Diagnostics and Design and Performance Validation of Phantoms Used in Conjunction with Optical Measurement of Tissue III*, vol. 7906, pp. 790613–790619, 10 February 2011.

[CAR 11b] CARDENAS N., YU L.F., MOHANTY S.K., "Stretching of red blood cells by optical tweezers quantified by digital holographic microscopy", *Optical Interactions with Tissue and Cells XXII*, vol. 7897, 2011.

[CAR 14] CARLSON N., *Physiology of Behavior*, Pearson, 2014.

[CHA 06a] CHARRIÈRE F., MARIAN A., MONTFORT F., *et al.*, "Cell refractive index tomography by digital holographic microscopy", *Optics Letters*, vol. 31, no. 2, pp. 178–180, 2006.

[CHA 06b] CHARRIÈRE F., PAVILLON N., COLOMB T., *et al.*, "Living specimen tomography by digital holographic microscopy: morphometry of testate amoeba", *Optics Express*, vol. 14, no. 16, pp. 7005–7013, 2006.

[CHO 09] CHOI Y.S., LEE S.J., "Three-dimensional volumetric measurement of red blood cell motion using digital holographic microscopy", *Applied Optics*, vol. 48, no. 16, pp. 2983–2990, 2009.

[CHU 13] CHUNG C.-Y., BOIK J., POTMA E.O., "Biomolecular imaging with coherent nonlinear vibrational microscopy", *Annual Review of Physical Chemistry*, vol. 64, no. 1, pp. 77–99, 2013.

[COG 97] COGSWELL C.J., SMITH N.I., LARKIN K.G., *et al.*, "Quantitative DIC microscopy using a geometric phase shifter", *Proceedings of Three-Dimensional Microscopy: Image Acquisition and Processing IV*, vol. 2984, pp. 72–81, 1997.

[COH 73] COHEN L.B., "Changes in neuron structure during action potential propagation and synaptic transmission", *Physiological Reviews*, vol. 53, no. 2, pp. 373–418, 1973.

[COL 10] COLOMB T., PAVILLON N., KUHN J., *et al.*, "Extended depth-of-focus by digital holographic microscopy", *Optics Letters*, vol. 35, no. 11, pp. 1840–1842, 2010.

[COR 14] CORNEA A., CONN P.M., *Fluorescence Microscopy: Super-Resolution and other Novel Techniques*, Academic Press, 2014.

[COS 03] COSSART R., ARONOV D., YUSTE R., "Attractor dynamics of network UP states in the neocortex", *Nature*, vol. 423, no. 6937, pp. 283–288, 2003.

[COT 13] COTTE Y., TOYM F., JOURDAIN P., *et al.*, "Marker-free phase nanoscopy (vol. 7, p. 113, 2013)", *Nature Photonics*, vol. 7, no. 5, pp. 418–418, 2013.

[CRH 11] CRHA I., ZAKOVA J., HUSER M., *et al.*, "Digital holographic microscopy in human sperm imaging", *Journal of Assisted Reproduction and Genetics*, vol. 28, no. 8, pp. 725–729, 2011.

[DAN 07] DANESHPANAH M., JAVIDI B., "Tracking biological microorganisms in sequence of 3D holographic microscopy images", *Optics Express*, vol. 15, no. 17, pp. 10761–10766, 2007.

[DAN 10] DANESHPANAH M., ZWICK S., SCHAAL F., *et al.*, "3D holographic imaging and trapping for non-invasive cell identification and tracking", *Journal of Display Technology*, vol. 6, no. 10, pp. 490–499, 2010.

[DAV 52] DAVIES H.G., WILKINS M.H.F., "Interference microscopy and mass determination", *Nature*, vol. 169, no. 4296, pp. 366–367, 1952.

[DUN 98] DUNN G.A., "Transmitted-light interference microscopy: a technique born before its time", *Proceedings of Royal Microscopical Society*, vol. 33, pp. 189–196, 1998.

[DUN 93] DUNN G.A., ZICHA D., "Phase-shifting interference microscopy applied to the analysis of cell behavior", *Cell Behaviour: Adhesion and Motility*, vol. 47, pp. 91–106, 1993.

[ELL 07] ELLERBEE A.K., CREAZZO T.L., IZATT J.A., "Investigating nanoscale cellular dynamics with cross-sectional spectral domain phase microscopy", *Optics Express*, vol. 15, no. 13, pp. 8115–8124, 2007.

[ESS 12] ESSELING M., KEMPER B., ANTKOWIAK M., *et al.*, "Multimodal biophotonic workstation for live cell analysis", *Journal of Biophotonics*, vol. 5, no. 1, pp. 9–13, 2012.

[FER 05] FERRARO P., GRILLI S., ALFIERI D., *et al.*, "Extended focused image in microscopy by digital holography", *Optics Express*, vol. 13, no. 18, pp. 6738–6749, 2005.

[FER 06] FERRARO P., GRILLI S., ALFIERI D., *et al.*, "3D imaging with large focus extension by a coherent optical microscope", in CRISTOBAL G., JAVIDI B., VALLMITJANA S. (eds.), *Information Optics, AIP Conf. Proc. 860*, vol. 860, pp. 52–62. Available at http://dx.doi.org/10.1063/1.2361 206, 2006.

[FRI 06] FRIEDRICH B., MATSKEVICH I., LANG F., "Cell volume regulatory mechanisms", *Contributions to Nephrology*, vol. 152, pp. 1–8, 2006.

[FU 10] FU D., OH S., CHOI W., *et al.*, "Quantitative DIC microscopy using an off-axis self-interference approach", *Optics Letters*, vol. 35, no. 14, pp. 2370–2372, 2010.

[FUJ 08] FUJITA K., SMITH N.I., "Label-free molecular imaging of living cells", *Molecules and Cells*, vol. 26, no. 6, pp. 530–535, 2008.

[GAO 11] GAO P., YAO B.L., HARDER I., *et al.*, "Phase-shifting Zernike phase contrast microscopy for quantitative phase measurement", *Optics Letters*, vol. 36, no. 21, pp. 4305–4307, 2011.

[GAR 10] GARAY J., D'ANGELO J.A., PARK Y., *et al.*, "Crosstalk between PKA and Epac regulates the phenotypic maturation and function of human dendritic cells", *Journal of Immunology*, vol. 185, no. 6, pp. 3227–3238, 2010.

[GHI 12] GHITA A., PASCUT F.C., MATHER M., *et al.*, "Cytoplasmic RNA in undifferentiated neural stem cells: a potential label-free Raman spectral marker for assessing the undifferentiated status", *Analytical Chemistry*, vol. 84, no. 7, pp. 3155–3162, 2012.

[HAM 05] HAMANN S., HERRERA-PEREZ J.J., BUNDGAARD M., *et al.*, "Water permeability of Na+-K+-2Cl- cotransporters in mammalian epithelial cells", *Journal of Physiology*, vol. 568, no. Pt 1, pp. 123–135, 2005.

[HAN 13] HAN R., LI Z., FAN Y., *et al.*, "Recent advances in super-resolution fluorescence imaging and its applications in biology", *Journal of Genetics and Genomics*, vol. 40, no. 12, pp. 583–595, 2013.

[HAY 12] HAYASAKI Y., SATO A., "Three-dimensional position measurement of a gold nanoparticle trapped in optical tweezers using in-line low-coherence digital holography", *Proceedings of the 2012 11th Euro-American Workshop on Information Optics (Wio)*, pp. 1–3, 2012.

[HIG 11] HIGUCHI T., PHAM Q.D., HASEGAWA S., *et al.*, "Three-dimensional positioning of optically trapped nanoparticles", *Applied Optics*, vol. 50, no. 34, pp. H183–H188, 2011.

[HIL 49] HILL D.K., KEYNES R.D., "Opacity changes in stimulated nerve", *Journal of Physiology*, vol. 108, no. 3, pp. 278–281, 1949.

[HOF 09] HOFFMANN E.K., LAMBERT I.H., PEDERSEN S.F., "Physiology of cell volume regulation in vertebrates", *Physiological Reviews*, vol. 89, no. 1, pp. 193–277, 2009.

[HOL 96] HOLTHOFF K., WITTE O.W., "Intrinsic optical signals in rat neocortical slices measured with near-infrared dark-field microscopy reveal changes in extracellular space", *Journal of Neuroscience*, vol. 16, no. 8, pp. 2740–2749, 1996.

[HSI 09] HSIEH C.L., GRANGE R., PU Y., *et al.*, "Three-dimensional harmonic holographic microcopy using nanoparticles as probes for cell imaging", *Optics Express*, vol. 17, no. 4, pp. 2880–2891, 2009.

[INO 97] INOUÉ S., SPRING R.K., *Video Microscopy, the Fundamentals*, Plenum, New York and London, 1997.

[ISH 99] ISHIWATA H., YATAGAI T., ITOH M., *et al.*, "Quantitative phase analysis in retardation modulated differential interference contrast (RM-DIC) microscope", *Proceedings of the Interferometry '99: Techniques and Technologies*, vol. 3744, pp. 183–187, 1999.

[JAN 09] JANECKOVA H., VESELY P., CHMELIK R., "Proving tumour cells by acute nutritional/energy deprivation as a survival threat: a task for microscopy", *Anticancer Research*, vol. 29, no. 6, pp. 2339–2345, 2009.

[JAN 12] JANG Y., JANG J., PARK Y., "Dynamic spectroscopic phase microscopy for quantifying hemoglobin concentration and dynamic membrane fluctuation in red blood cells", *Optics Express*, vol. 20, no. 9, pp. 9673–9681, 2012.

[JAV 10a] JAVIDI B., DANESHPANAH M., MOON I., "Three-dimensional holographic imaging for identification of biological micro/nanoorganisms", *IEEE Photonics Journal*, vol. 2, no. 2, pp. 256–259, 2010.

[JAV 10b] JAVIDI B., MOON I., DANESHPANAHA M., "Detection, identification and tracking of biological micro/nano organisms by computational 3D optical", *II, Proc. SPIE 7759*, vol. 7759, 24 August 2010.

[JOU 11] JOURDAIN P., PAVILLON N., MORATAL C., *et al.*, "Determination of transmembrane water fluxes in neurons elicited by glutamate ionotropic receptors and by the cotransporters KCC2 and NKCC1: a digital holographic microscopy study", *Journal of Neuroscience*, vol. 31, no. 33, pp. 11846–11854, 2011.

[JOU 12] JOURDAIN P., BOSS D., RAPPAZ B., *et al.*, "Simultaneous optical recording in multiple cells by digital holographic microscopy of chloride current associated to activation of the ligand-gated chloride channel GABA(A) receptor", *Plos One*, vol. 7, no. 12, p. e51041, 2012.

[JOU 14] JOURDAIN P., BECQ F., LENGACHER S., *et al.*, "The human CFTR protein expressed in CHO cells activates aquaporin-3 in a CAMP-dependent pathway: study by digital holographic microscopy", *Journal of Cell Science*, vol. 127, no. Pt 3, pp. 546–556, 2014.

[KEM 07] KEMPER B., KOSMEIER S., LANGEHANENBERG P., *et al.*, "Integral refractive index determination of living suspension cells by multifocus digital holographic phase contrast microscopy", *Journal of Biomedical Optics*, vol. 12, no. 5, pp. 1–5, 19 October 2007.

[KEM 10] KEMPER B., BAUWENS A., VOLLMER A., *et al.*, "Label-free quantitative cell division monitoring of endothelial cells by digital holographic microscopy", *Journal of Biomedical Optics*, vol. 15, no. 3, p. 036009, 2010.

[KEM 13] KEMPER B., BARROSO A., WOERDEMANN M., *et al.*, "Towards 3D modeling and imaging of infection scenarios at the single cell level using holographic optical tweezers and digital holographic microscopy", *Journal of Biophotonics*, vol. 6, no. 3, pp. 260–266, 2013.

[KIM 09] KIM T., POON T.C., "Autofocusing in optical scanning holography", *Applied Optics*, vol. 48, no. 34, pp. H153–H159, 2009.

[KIM 12] KIM M., CHOI Y., FANG-YEN C., *et al.*, "Three-dimensional differential interference contrast microscopy using synthetic aperture imaging", *Journal of Biomedical Optics*, vol. 17, no. 2, pp. 1–6, 23 February 2012.

[KIN 08] KING S.V., LIBERTUN A., PIESTUN R., *et al.*, "Quantitative phase microscopy through differential interference imaging", *Journal of Biomedical Optics*, vol. 13, no. 2, pp. 1–10, 21 April 2008.

[KOU 10] KOU S.S., WALLER L., BARBASTATHIS G., *et al.*, "Transport-of-intensity approach to differential interference contrast (TI-DIC) microscopy for quantitative phase imaging", *Optics Letters*, vol. 35, no. 3, pp. 447–449, 2010.

[KUH 07] KUHN J., COLOMB T., MONTFORT F., *et al.*, "Real-time dual-wavelength digital holographic microscopy with a single hologram acquisition", *Optics Express*, vol. 15, no. 12, pp. 7231–7242, 2007.

[KUH 13] KUHN J., SHAFFER E., MENA J., *et al.*, "Label-free cytotoxicity screening assay by digital holographic microscopy", *Assay and Drug Development Technologies*, vol. 11, no. 2, pp. 101–107, 2013.

[KUM 14] KUMAR V., ABBAS A.K., ASTER J., *Robbins & Cotran Pathologic Basis of Disease*, Elsevier Saunders, 2014.

[KUS 14] KUS A., DUDEK M., KEMPER B., *et al.*, "Tomographic phase microscopy of living three-dimensional cell cultures", *Journal of Biomedical Optics*, vol. 19, no. 4, pp. 046009, 2014.

[LAN 98] LANG F., BUSCH G.L., RITTER M., *et al.*, "Functional significance of cell volume regulatory mechanisms", *Physiological Reviews*, vol. 78, no. 1, pp. 247–306, 1998.

[LAN 09] LANGEHANENBERG P., IVANOVA L., BERNHARDT I., *et al.*, "Automated three-dimensional tracking of living cells by digital holographic microscopy", *Journal of Biomedical Optics*, vol. 14, no. 1, pp. 014018, 2009.

[LEE 11] LEE S., LEE J.Y., PARK C.S., *et al.*, "Detrended fluctuation analysis of membrane flickering in discocyte and spherocyte red blood cells using quantitative phase microscopy", *Journal of Biomedical Optics*, vol. 16, no. 7, p. 076009, 2011.

[LEE 12] LEE K.F.H., SOARES C., BEIQUE J.C., "Examining form and function of dendritic spines", *Neural Plasticity*, vol. 12, no. 704103, p. 9 2012.

[LEE 13] LEE K., KIM K., JUNG J., *et al.*, "Quantitative phase imaging techniques for the study of cell pathophysiology: from principles to applications", *Sensors*, vol. 13, no. 4, pp. 4170–4191, 2013.

[LIE 04] LIEBLING M., UNSER M., "Autofocus for digital Fresnel holograms by use of a Fresnelet-sparsity criterion", *Journal of the Optical Society of America a-Optics Image Science and Vision*, vol. 21, no. 12, pp. 2424–2430, 2004.

[LIU 11] LIU R., DEY D.K., BOSS D., *et al.*, "Recognition and classification of red blood cells using digital holographic microscopy and data clustering with discriminant analysis", *Journal of the Optical Society of America a-Optics Image Science and Vision*, vol. 28, no. 6, pp. 1204–1210, 2011.

[LUE 06] LUE N., POPESCU G., IKEDA T., *et al.*, "Live cell refractometry using microfluidic devices", *Optics Letters*, vol. 31, no. 18, pp. 2759–2761, 2006.

[MAC 91] MACVICAR B.A., HOCHMAN D., "Imaging of synaptically evoked intrinsic optical signals in hippocampal slices", *Journal of Neuroscience*, vol. 11, no. 5, pp. 1458–1469, 1991.

[MAN 05] MANN C.J., YU L.F., LO C.M., *et al.*, "High-resolution quantitative phase-contrast microscopy by digital holography", *Optics Express*, vol. 13, no. 22, pp. 8693–8698, 2005.

[MAN 06] MANN C.J., YU L.F., KIM M.K., "Movies of cellular and sub-cellular motion by digital holographic microscopy", *Biomedical Engineering Online*, vol. 5, p. 21, 2006.

[MAR 05] MARQUET P., RAPPAZ B., MAGISTRETTI P.J., *et al.*, "Digital holographic microscopy: a non-invasive contrast imaging technique allowing quantitative visualization of living cells with subwavelength axial accuracy", *Optics Letters*, vol. 30, no. 5, pp. 468–470, 2005.

[MAR 13] MARQUET P., DEPEURSINGE C., MAGISTRETTI P.J., "Exploring neural cell dynamics with digital holographic microscopy", *Annual Review of Biomedical Engineering*, vol. 15, pp. 407–431, 2013.

[MAU 10] MAURITZ J.M.A., ESPOSITO A., TIFFERT T., *et al.*, "Biophotonic techniques for the study of malaria-infected red blood cells", *Medical & Biological Engineering & Computing*, vol. 48, no. 10, pp. 1055–1063, 2010.

[MCE 08] MCELHINNEY C., HENNELLY B., NAUGHTON T., "Extended focused imaging for digital holograms of macroscopic three-dimensional objects", *Applied Optics*, vol. 47, no. 19, pp. D71–D79, 2008.

[MEH 12] MEHTA D.S., SRIVASTAVA V., "Quantitative phase imaging of human red blood cells using phase-shifting white light interference microscopy with colour fringe analysis", *Applied Physics Letters*, vol. 101, no. 20, p. 203701, 2012.

[MEM 11] MEMMOLO P., DI CAPRIO G., DISTANTE C., *et al.*, "Identification of bovine sperm head for morphometry analysis in quantitative phase-contrast holographic microscopy", *Optics Express*, vol. 19, no. 23, pp. 23215–23226, 2011.

[MEM 12] MEMMOLO P., IANNONE M., VENTRE M., *et al.*, "On the holographic 3D tracking of *in vitro* cells characterized by a highly-morphological change", *Optics Express*, vol. 20, no. 27, pp. 28485–28493, 2012.

[MER 12] MEROLA F., MICCIO L., MEMMOLO P., *et al.*, "Simultaneous optical manipulation, 3-D tracking, and imaging of micro-objects by digital holography in microfluidics", *IEEE Photonics Journal*, vol. 4, no. 2, pp. 451–454, 2012.

[MER 13] MEROLA F., MICCIO L., MEMMOLO P., *et al.*, "Digital holography as a method for 3D imaging and estimating the biovolume of motile cells", *Lab on a Chip*, vol. 13, no. 23, pp. 4512–4516, 2013.

[MIC 11] MICCIO L., FINIZIO A., MEMMOLO P., *et al.*, "Detection and visualization improvement of spermatozoa cells by digital holography", in LIN C.P., NTZIACHRISTOS V. (eds.), *Molecular Imaging III, Proceedings of SPIE*, vol. 8089, pp. 1605–7422, 2011.

[MIH 11] MIHAILESCU M., SCARLAT M., GHEORGHIU A., *et al.*, "Automated imaging, identification, and counting of similar cells from digital hologram reconstructions", *Applied Optics*, vol. 50, no. 20, pp. 3589–3597, 2011.

[MIL 02] MILGRAM J.H., LI W.C., "Computational reconstruction of images from holograms", *Applied Optics*, vol. 41, no. 5, pp. 853–864, 2002.

[MIR 11a] MIR M., TANGELLA K., POPESCU G., "Blood testing at the single cell level using quantitative phase and amplitude microscopy", *Biomedical Optics Express*, vol. 2, no. 12, pp. 3259–3266, 2011.

[MIR 11b] MIR M., WANG Z., SHEN Z., *et al.*, "Optical measurement of cycle-dependent cell growth", *Proceedings of the National Academy of Sciences of the United States of America*, vol. 108, no. 32, pp. 13124–13129, 2011.

[MIT 56] MITCHISON J.M., PASSANO L.M., SMITH F.H., "An integration method for the interference microscope", *Quarterly Journal of Microscopical Science*, vol. 97, no. 2, pp. 287–302, 1956.

[MOL 08] MOLDER A., SEBESTA M., GUSTAFSSON M., *et al.*, "Non-invasive, label-free cell counting and quantitative analysis of adherent cells using digital holography", *Journal of Microscopy*, vol. 232, no. 2, pp. 240–247, 2008.

[MOO 07] MOON I., JAVIDI B., "Three-dimensional identification of stem cells by computational holographic imaging", *Journal of the Royal Society Interface*, vol. 4, no. 13, pp. 305–313, 2007.

[MOO 10] MOON I., YI F., JAVIDI B., "Automated three-dimensional microbial sensing and recognition using digital holography and statistical sampling", *Sensors*, vol. 10, no. 9, pp. 8437–8451, 2010.

[MOO 12] MOON I., JAVIDI B., YI F., *et al.*, "Automated statistical quantification of three-dimensional morphology and mean corpuscular hemoglobin of multiple red blood cells", *Optics Express*, vol. 20, no. 9, pp. 10295–10309, 2012.

[MOO 13] MOON I., YI F., LEE Y.H., *et al.*, "Automated quantitative analysis of 3D morphology and mean corpuscular hemoglobin in human red blood cells stored in different periods", *Optics Express*, vol. 21, no. 25, pp. 30947–30957, 2013.

[NAG 04] NAGELHUS E.A., MATHIISEN T.M., OTTERSEN O.P., "Aquaporin-4 in the central nervous system: cellular and subcellular distribution and coexpression with KIR4.1", *Neuroscience*, vol. 129, no. 4, pp. 905–913, 2004.

[NOM 50] NOMARSKI G., "Dispositif oculaire à contraste de phase pour microscope", *Journal De Physique Et Le Radium*, vol. 11, no. 12, pp. S9–S10, 1950.

[OKA 12] OKADA M., SMITH N.I., PALONPON A.F., *et al.*, "Label-free Raman observation of cytochrome c dynamics during apoptosis", *Proceedings of the National Academy of Sciences of the United States of America*, vol. 109, no. 1, pp. 28–32, 2012.

[PAD 11] PADAMSEY Z., EMPTAGE N.J., "Imaging synaptic plasticity", *Molecular Brain*, vol. 4, no. 36, pp. 1–20, 29 September 2011.

[PAN 13] PAN W.Q., "Multiplane imaging and depth-of-focus extending in digital holography by a single-shot digital hologram", *Optics Communications*, vol. 286, pp. 117–122, 2013.

[PAR 06] PARK Y., POPESCU G., BADIZADEGAN K., *et al.*, "Diffraction phase and fluorescence microscopy", *Optics Express*, vol. 14, no. 18, pp. 8263–8268, 2006.

[PAR 09] PARK Y., YAMAUCHI T., CHOI W., *et al.*, "Spectroscopic phase microscopy for quantifying hemoglobin concentrations in intact red blood cells", *Optics Letters*, vol. 34, no. 23, pp. 3668–3670, 2009.

[PAR 10b] PARK Y., DIEZ-SILVA M., FU D., *et al.*, "Static and dynamic light scattering of healthy and malaria-parasite invaded red blood cells", *Journal of Biomedical Optics*, vol. 15, no. 2, p. 020506, 2010.

[PAR 10b] PARK Y., BEST C.A., AUTH T., *et al.*, "Metabolic remodeling of the human red blood cell membrane", *Proceedings of the National Academy of Sciences of the United States of America*, vol. 107, no. 4, pp. 1289–1294, 2010.

[PAR 11] PARK Y., BEST C.A., KURIABOVA T., *et al.*, "Measurement of the nonlinear elasticity of red blood cell membranes", *Physical Review E*, vol. 83, no. 5, pp. 1–5, 27 May 2011.

[PAS 10] PASANTES-MORALES H., CRUZ-RANGEL S., "Brain volume regulation: osmolytes and aquaporin perspectives", *Neuroscience*, vol. 168, no. 4, pp. 871–884, 2010.

[PAT 12] PATURZO M., FINIZIO A., MEMMOLO P., *et al.*, "Microscopy imaging and quantitative phase contrast mapping in turbid microfluidic channels by digital holography", *Lab on a Chip*, vol. 12, no. 17, pp. 3073–3076, 2012.

[PAV 10] PAVILLON N., BENKE A., BOSS D., *et al.*, "Cell morphology and intracellular ionic homeostasis explored with a multimodal approach combining epifluorescence and digital holographic microscopy", *Journal of Biophotonics*, vol. 3, no. 7, pp. 432–436, 2010.

[PAV 12] PAVILLON N., KUHN J., MORATAL C., *et al.*, "Early cell death detection with digital holographic microscopy", *Plos One*, vol. 7, no. 1, p. e30912, 2012.

[PAV 13] PAVILLON N., HOBRO A.J., SMITH N.I., "Cell optical density and molecular composition revealed by simultaneous multimodal label-free imaging", *Biophysical Journal*, vol. 105, no. 5, pp. 1123–1132, 2013.

[PAV 14] PAVILLON N., FUJITA K., SMITH N.I., "Multimodal label-free microscopy", *Journal of Innovative Optical Health Sciences*, vol. 7, no. 5, pp. 1330009–1330009, 2014.

[PET 10] PETIBOIS C., "Imaging methods for elemental, chemical, molecular, and morphological analyses of single cells", *Analytical and Bioanalytical Chemistry*, vol. 397, no. 6, pp. 2051–2065, 2010.

[PET 11] PETERKA D.S., TAKAHASHI H., YUSTE R., “Imaging voltage in neurons”, *Neuron*, vol. 69, no. 1, pp. 1–5, 27 May 2011.

[PHA 12] PHAM H., BHADURI B., DING H.F., *et al.*, “Spectroscopic diffraction phase microscopy”, *Optics Letters*, vol. 37, no. 16, pp. 3438–3440, 2012.

[POP 05] POPESCU G., IKEDA T., BEST C.A., *et al.*, “Erythrocyte structure and dynamics quantified by Hilbert phase microscopy”, *Journal of Biomedical Optics*, vol. 10, no. 6, pp. 1–3, 30 December 2005.

[POP 08] POPESCU G., “Quantitative phase imaging of nanoscale cell structure and dynamics”, *Methods in Nano Cell Biology*, vol. 90, p. 87, 2008.

[POP 08a] POPESCU G., PARK Y., CHOI W., *et al.*, “Imaging red blood cell dynamics by quantitative phase microscopy”, *Blood Cells Molecules and Diseases*, vol. 41, no. 1, pp. 10–16, 2008.

[POP 08b] POPESCU G., PARK Y., LUE N., *et al.*, “Optical imaging of cell mass and growth dynamics”, *American Journal of Physiology-Cell Physiology*, vol. 295, no. 2, pp. C538–C544, 2008.

[RAP 08] RAPPAZ B., BARBUL A., EMERY Y., *et al.*, “Comparative study of human erythrocytes by digital holographic microscopy, confocal microscopy, and impedance volume analyzer”, *Cytometry Part A*, vol. 73A, no. 10, pp. 895–903, 2008.

[RAP 09] RAPPAZ B., BARBUL A., HOFFMANN A., *et al.*, “Spatial analysis of erythrocyte membrane fluctuations by digital holographic microscopy”, *Blood Cells Molecules and Diseases*, vol. 14, no. 3, pp. 1–5, 16 June 2009.

[RAP 14] RAPPAZ B., BRETON B., SHAFFER E., *et al.*, “Digital holographic microscopy: a quantitative label-free microscopy technique for phenotypic screening”, *Combinatorial Chemistry & High Throughput Screening*, vol. 17, no. 1, pp. 80–88, 2014.

[RAP 05] RAPPAZ B., MARQUET P., CUCHE E., *et al.*, “Measurement of the integral refractive index and dynamic cell morphometry of living cells with digital holographic microscopy”, *Optics Express*, vol. 13, no. 23, pp. 9361–9373, 2005.

[RAP 08] RAPPAZ B., CHARRIÈRE F., DEPEURSINGE C., *et al.*, “Simultaneous cell morphometry and refractive index measurement with dual-wavelength digital holographic microscopy and dye-enhanced dispersion of perfusion medium”, *Optics Letters*, vol. 33, no. 7, pp. 744–746, 2008.

[RAP 09] RAPPAZ B., CANO E., COLOMB T., *et al.*, "Non-invasive characterization of the fission yeast cell cycle by monitoring dry mass with digital holographic microscopy", *Journal of Biomedical Optics*, vol. 14, no. 3, 2009.

[RIN 12] RINEHART M., ZHU Y.Z., WAX A., "Quantitative phase spectroscopy", *Biomedical Optics Express*, vol. 3, no. 5, pp. 958–965, 2012.

[SEO 10] SEO S., ISIKMAN S.O., SENCAN I., *et al.*, "High-throughput lens-free blood analysis on a chip", *Analytical Chemistry*, vol. 82, no. 11, pp. 4621–4627, 2010.

[SHA 10] SHAFFER E., MARQUET P., DEPEURSINGE C., "Real time, nanometric 3D-tracking of nanoparticles made possible by second harmonic generation digital holographic microscopy", *Optics Express*, vol. 18, no. 16, pp. 17392–17403, 2010.

[SHA 12] SHAKED N.T., SATTERWHITE L.L., TELEN M.J., *et al.*, "Dynamic quantitative microscopy and nanoscopy of red blood cells in sickle cell disease", *Three-Dimensional and Multidimensional Microscopy: Image Acquisition and Processing XIX, Proc. SPIE 8227*, vol. 8227, 9 February 2012.

[SIM 10] SIMON B., DEBAILLEUL M., BEGHIN A., *et al.*, "High-resolution tomographic diffractive microscopy of biological samples", *Journal of Biophotonics*, vol. 3, no. 7, pp. 462–467, 2010.

[STE 91] STEPNOSKI R.A., LAPORTA A., RACCUIA-BEHLING F., *et al.*, "Non-invasive detection of changes in membrane potential in cultured neurons by light scattering", *Proceedings of the National Academy of Sciences of the United States of America*, vol. 88, no. 21, pp. 9382–9386, 1991.

[STR 04] STRANGE K., "Cellular volume homeostasis", *Advances in Physiology Education*, vol. 28, nos. 1–4, pp. 155–159, 2004.

[SUN 08] SUN H.Y., SONG B., DONG H.P., *et al.*, "Visualization of fast-moving cells *in vivo* using digital holographic video microscopy", *Journal of Biomedical Optics*, vol. 13, no. 1, pp. 1–9, 19 February 2008.

[SVO 06] SVOBODA K., YASUDA R., "Principles of two-photon excitation microscopy and its applications to neuroscience", *Neuron*, vol. 50, no. 6, pp. 823–839, 2006.

[SYK 03] SYKOVÀ E., VARGOVÀ L., KUBINOVÀ S., *et al.*, "The relationship between changes in intrinsic optical signals and cell swelling in rat spinal cord slices", *Neuroimage*, vol. 18, no. 2, pp. 214–230, 2003.

[TAS 68] TASAKI I., WATANABE A., SANDLIN R., *et al.*, "Changes in fluorescence, turbidity, and birefringence associated with nerve excitation", *Proceedings of the National Academy of Sciences of the United States of America*, vol. 61, no. 3, pp. 883–888, 1968.

[TAS 92] TASAKI I., BYRNE P.M., "Rapid structural changes in nerve fibers evoked by electric current pulses", *Biochemical and Biophysical Research Communications*, vol. 188, no. 2, pp. 559–564, 1992.

[TOY 12] TOY M.F., RICHARD S., KUHN J., *et al.*, "Digital holographic microscopy for the cytomorphological imaging of cells under zero gravity", *Three-Dimensional and Multidimensional Microscopy: Image Acquisition and Processing XIX, Proc. SPIE*, vol. 8227, 9 February 2012.

[WAN 09] WANG H.Y., QIN A.L., HUANG M., *et al.*, "Autofocus method for digital holographic reconstruction of microscopic object", *IEEE Photonics and Optoelectronics*, pp. 1–4, 14–16 August 2009.

[WAN 10] WANG P., BISTA R.K., KHALBUSS W.E., *et al.*, "Nanoscale nuclear architecture for cancer diagnosis beyond pathology via spatial-domain low-coherence quantitative phase microscopy", *Journal of Biomedical Optics*, vol. 15, no. 6, p. 066028, 2010.

[WAN 11] WANG P., BISTA R.K., KHALBUSS W.E., *et al.*, "Nanoscale nuclear architecture for cancer diagnosis by spatial-domain low-coherence quantitative phase microscopy", *Biomedical Applications of Light Scattering V*, vol. 7907, pp. 1–8, 11 February 2011.

[WAR 10] WARNASOORIYA N., JOUD F., BUN P., *et al.*, "Imaging gold nanoparticles in living cell environments using heterodyne digital holographic microscopy", *Optics Express*, vol. 18, no. 4, pp. 3264–3273, 2010.

[WU 11] WU S.M., HOCHEDLINGER K., "Harnessing the potential of induced pluripotent stem cells for regenerative medicine (vol. 13, p. 497, 2011)", *Nature Cell Biology*, vol. 13, no. 6, pp. 734–734, 2011.

[XU 03] XU W., JERICHO M.H., KREUZER H.J., *et al.*, "Tracking particles in four dimensions with in-line holographic microscopy", *Optics Letters*, vol. 28, no. 3, pp. 164–166, 2003.

[YI 12] YI F., LEE C.G., MOON I.K., "Statistical analysis of 3D volume of red blood cells with different shapes via digital holographic microscopy", *Journal of the Optical Society of Korea*, vol. 16, no. 2, pp. 115–120, 2012.

[YI 13] YI F., MOON I., JAVIDI B., *et al.*, "Automated segmentation of multiple red blood cells with digital holographic microscopy", *Journal of Biomedical Optics*, vol. 18, no. 2, p. 26006, 2013.

[YIN 12] YIN Z.Z., KANADE T., CHEN M., "Understanding the phase contrast optics to restore artifact-free microscopy images for segmentation", *Medical Image Analysis*, vol. 16, no. 5, pp. 1047–1062, 2012.

[YUS 09] YUSIPOVICH A.I., PARSHINA E.Y., BRYSGALOVA N.Y., *et al.*, "Laser interference microscopy in erythrocyte study", *Journal of Applied Physics*, vol. 105, no. 10, pp. 102037, 2009.

[ZER 42] ZERNIKE F., "Phase contrast, a new method for the microsopic observation of transparent objects", *Physica*, vol. 9, pp. 686–698, 1942.

[ZHA 10] ZHAO J., WANG D.Y., WANG Y.X., *et al.*, "Autofocusing on pure phase object for living cell imaging in lensless Fourier transform digital holography", in TOWERS C.E., SCHMIT J., CREATH K. (eds.), *Interferometry XV: Techniques and Analysis Proc. SPIE 7790*, vol. 779017, 2 August 2010.

[ZIC 95] ZICHA D., DUNN G.A., "An image-processing system for cell behavior studies in subconfluent cultures", *Journal of Microscopy-Oxford*, vol. 179, pp. 11–21, 1995.

6

Long-Wave Infrared Digital Holography

6.1. Introduction

Usually, holography (both analog and digital) is applied in visible light. This can be understood simply because people like to see their beams when they align optical setups. However, once the operator has passed the psychological barrier of invisible beams, holography at non-visible wavelengths has proven to uncover really exciting features.

On the shorter wavelength side, the use of ultraviolet (UV) in holographic setups was proposed to study micrometric devices [ASW 03a, ASW 03b]. The significance of shorter wavelengths is to increase imaging resolution which otherwise is reduced by diffraction if we use visible wavelengths. Another advantage is to reduce sensitivity to displacements which are proportional to the wavelength, as can be seen in equation [1.68] in Chapter 1. Finally, as the scattering properties of surfaces are directly related to wavelength, scattering appears more easily on polished or low-roughness surfaces at short wavelengths [YAM 03]. In particular, reflection of micromirror structures induces speckle at 266 nm wavelengths,

---

Chapter written by Marc GEORGES.

rendering possible the development of electronic speckle pattern interferometry on such structures [ASW 03a, ASW 03b]. Shorter UV wavelengths (193 nm) were used more recently in digital holographic microscopy (DHM) devices with various configurations showing excellent resolutions on microstructures and biological samples [PED 07, FAR 10]. Deeper in wavelengths, X-ray holography was also demonstrated and shows very high potential for nanometric object studies [EIS 04, DUC 11].

Concerning the longer wavelengths, before going further, must we first clarify the terminology which we use in this chapter. The infrared (IR) spectrum, extending from 700 nm to 1 mm wavelengths, is divided into several bands and some terms may designate different bands, depending on classification diagrams. We will not refer to the classifications recommended by the International Organization for Standardization (ISO) (norm ISO 20473) or the International Commission on Illumination (ICE), which both divide the spectrum into three bands. Rather, we prefer a more common classification which is well adopted by the community at large: the near infrared (NIR) extending from 700 nm to 1 µm, the short-wave infrared (SWIR) extending from 1 to 3 µm, the mid-wave infrared (MWIR) extending from 3 to 8 µm, the long-wave infrared (LWIR) extending from 8 to 15 µm and, finally, the far infrared (FIR) extending from 15 µm to 1 mm. The NIR and SWIR are sometimes referred to as reflecting IR bands, the MWIR and LWIR are referred to as the thermal IR bands, and the FIR is often called the terahertz domain and constitutes the limit with the millimeter/microwave domain.

Holography has been applied in most of the above mentioned IR bands. As for any wavelength, it is sufficient to have a laser, optical components which form the beams and the image, an imaging device sensitive to the laser wavelength and a holographic recording medium. In digital holography, we suppose the latter is the array sensor of the imaging device.

NIR and SWIR digital holographic systems can be seen as adaptation of the visible systems because the optical components are essentially the same (silica or borosilicate glasses are still transparent), the only differences being the laser sources and the detectors. The SWIR digital holography experiments reported so far worked in transmission under various configurations. Despite the loss of resolution due to diffraction compared to visible wavelengths, the SWIR domain facilitates inspecting some materials which are opaque in visible but transparent in SWIR. Repetto *et al.* showed an in-line Gabor setup for inspecting aluminum bolometers deposited on silicon wafers [REP 05]. They used a superluminescent diode at 1.3 μm. Delacretaz *et al.* used a digital holographic microscope based on a Mach–Zehnder diagram with a 1.55 μm tunable distributed feedback laser used in telecom [DEL 09] or 1.28 μm [DEL 10]. Ryle *et al.* used a Gabor configuration with a tunable 1.55 μm laser to demonstrate recordings at two distinct but close wavelengths for performing the two-wavelength holography technique [RYL 11]. All adopted configurations facilitate using such laser sources with relatively low coherence lengths.

In this chapter, we will present what has been achieved so far in holography in the LWIR domain, covering mainly the use of CO_2 lasers around the 10 μm wavelength. First, we will review the early works based on analog recording. Then, we will present the digital recording of holograms in LWIR, first by reviewing the main hardware components, and discuss some specific features of using such long wavelength. Afterward, we will present applications which take advantages of the specific features of long-wave infrared digital holography (LWIR DH). We will see some true innovative aspects which were demonstrated and which are not possible if a visible wavelength was used.

6.2. Analog hologram recording in LWIR

Holography with CO_2 lasers is not a new subject since it was first shown in 1969 by Chivian *et al.* who used a thermochromic material (cuprous mercuric iodide) for recording at 10.6 μm and read-out in the visible with a HeNe laser [CHI 69]. Many works followed by other groups which used a similar setup and recording medium [ROB 76] or other recording media: liquid crystals [SIM 70], wax and gelatin films

[KOB 71], bismuth thin films [DEC 72, FOR 73, ROB 76], acrylic and thin films [RIO 77], wax [BEA 77, BEA 78], oil films deposited on glass plates [LEW 84], resists [BEA 94], poly(acryl acid) films [BEA 00] and albumen [CAL 03]. All these materials are able to record patterns through relief variations, producing phase holograms that can be processed *in situ*. Also, they show relatively good figures of merit in terms of diffraction efficiency. But what seems to remain limited is the resolution of such media with a rapid drop of diffraction efficiency for line spacings larger than 10 lines per millimeter. Also, they all use visible wavelengths for readout which does not make easy the application of these materials for real-time holographic interferometry (HI). Double-exposure holograms which are read-out with visible lasers have also been tried [DEC 72, LEW 86].

6.3. Digital hologram recording in LWIR

6.3.1. *Hardware components*

6.3.1.1. *Lasers*

The most common coherent source in the LWIR domain is the CO_2 laser. The technology is robust and proved and the industry makes large use of it in various material processing applications, such as cutting, welding and marking, or for medical purposes. Pulse and continuous modes can be found with power ranging from gigawatts to milliwatts. The emission lines depend on the isotopes of carbon and oxide [CHE 87]. However, the most common ones are 10.6, 10.3, 9.6 and 9.3 μm, depending on the manufacturer and models. In the case of holography, the CO_2 lasers, which are currently used, have power ranging from a few watts to a few tens of watts. The coherence lengths are very long. The latter is directly related to the emission linewidth $\Delta\nu_{line}$, which itself depends on the pressure of the gas mixture [SVE 84] and can be different from one model of laser to another. Generally such CO_2 lasers can have linewidths of a few tens of kHz which ensures coherence lengths larger than hundred meters which is highly comfortable for holography. However, if we wish to apply LWIR holography for holographic interferometry purposes, we need a CO_2 laser with optional stabilization which is proposed by some manufacturers. Indeed, CO_2 lasers usually suffer from mode hop which consists of change of longitudinal mode which has the effect of changing wavelength. If we record two holograms, one before and one after the mode hop, the difference of phases calculated by any DH principle will show fringes related to optical path difference due to the wavelength change. This can be used advantageously for holographic contouring of a stable object [KRE 05] but is a problem if we wish to apply digital holographic interferometry (DHI) for observing the movement or deformation of an object, for which we need a stable wavelength during the measurement. To the best of our knowledge, only Edinburgh Instruments and Access Laser Company propose this stabilization for CO_2 lasers typically emitting 10 W, which is sufficient for most LWIR digital holographic interferometers presented in this chapter. The company VM-TIM manufactured also stabilized CO_2 lasers, and which were used in some LWIR DHI works [ALE 13], but recently it discarded this product.

The quantum cascade laser technology offers the possibility of laser lines between 3 and 25 µm, moreover with the possibility of tuning the wavelength [YAO 12]. Ravaro *et al.* [RAV 14] recently demonstrated applications of such lasers in various DH applications. In particular they showed two-wavelength DHI which can be used for contouring [KRE 05].

6.3.1.2. *Camera technologies*

The two important classes of IR focal plane arrays (FPAs), which are applied in LWIR imaging, are based on photon detection or thermal detection [KRU 01]. We will briefly review the technologies which are used in interferometric applications.

The photon detection class is mostly represented in LWIR by two technologies. The first technology is the quantum well infrared photodetectors (QWIPs) [SCH 07, LEV 93]. They are made up of III–V semiconductor materials containing one or more quantum wells formed by layers of different materials with different bandgaps. Photons are absorbed through transitions between the fundamental electronic level of the quantum well and its first excited state. The absorption is resonant, yielding a narrow spectral sensitivity bandwidth. The spectral sensitivity can be tailored by careful design of the quantum well parameters (materials, layers, etc.). Generally, the spectral sensitivity of QWIPs is in the range of 8–9 μm with a bandwidth of 1 μm but we can find examples in literature of sensitivities centered at 10 μm wavelengths [DIA 07], making them compatible with some lines of the CO_2 laser. The second technology is the one of mercury cadmium telluride (MCT) made up of alloys of II–VI semiconductor materials, HgTe and CdTe, with various stoichiometries [ROG 05]. The MCT technology can cover SWIR, MWIR and LWIR ranges, or even higher, depending on the alloy $Hg_{1-x}Cd_xTe$ chosen. In the LWIR range, the spectral sensitivity generally extends between 8 and 10 μm with strong cutoff at the highest wavelength, making the MCTs compatible with some CO_2 laser lines. State-of-the-art LWIR FPAs can be found in 640 × 512 pixels format with 15–20 μm pitch (depending on manufacturer) while in the MWIR range, 1,280 × 1,024 pixels FPAs are available. Recently, 5 μm pitches were demonstrated [ARM 14]. This shows that the MCT technology is in a constant and promising evolution in terms of performance and compactness.

In order to show optimum detection performances, QWIP and MCT technologies require cryogenic cooling operation. Indeed, in addition to the photoexcited carriers, generated by photon absorption, a certain number of thermally excited carriers exist and their density is too high at room temperature, causing important noise. Therefore, for obtaining optimum performances, FPAs are cooled generally by active Stirling coolers which have the problem of inducing vibrations to the sensor. Although these vibrations have a negligible impact on classical thermal imaging, it strongly prevents the use of such imagers in holography. Indeed, a usual stability requirement of a holographic setup is that the interference between object and reference beams must not move of a fraction of wavelength during the hologram capture. It is then obvious that the FPA has to be sufficiently stable for holography. An LWIR camera manufactured by Infratec (model ImageIR 8800) was developed for non-destructive testing [JOR 13]. It is based on LWIR Hawk MCT FPA made by Selex company. We have tried to use such camera for holography but we did not succeed so far due to the cryocooler vibrations.

The thermal detection class of FPAs is the main competitor of the former since it does not require cooling; hence they are often referred to as the uncooled IR detectors. The main uncooled technologies used in LWIR imaging are the pyroelectric and the microbolometer arrays [KRU 01].

One of the most used effects for IR detection is the pyroelectric effect. This effect has been exploited mostly in ferroelectrics: it exhibits spontaneous electric polarization which yields a measurable voltage at electrodes placed at opposite faces of the material. There is no voltage difference between electrodes when temperature does not change, because polarization is neutralized by mobile charges on the sample surface. When temperature changes the electric polarization also changes, hence the voltage difference is further neutralized by moving charges. Therefore, pyroelectric detectors have no direct current (DC) signals, but instead measure temperature varying in time [KRU 01].

The earliest technology used for LWIR imaging was that of pyroelectric vidicon (PEV) [CON 76]. The vidicon technology was developed in the 1950s to record refreshable images electronically. A photoconductor material deposited on the internal face of the transparent window of a cathodic tube serves as a target, which accumulates charges due to the incident light coming from a scene through a lens. This charge pattern is scanned by a low-velocity electron beam whose voltage is monitored during the scanning and amplified to reproduce the scene imaged on the photoconductive target. The PEV principle is the same except that the target is made of pyroelectric material. As a result, the drawback of PEV is its sensitivity to temperature variations of the scene. Depending on the window material sealing the tube, PEVs could image between 2 and 400 μm, hence covering the LWIR range. PEVs were used in the very first commercial 10.6 μm interferometer which required varying the

IR light intensity by introducing a phase shift in one arm of the interferometer [KWO 80]. PEV also allowed the very first implementation of electronic hologram recording in the LWIR [LØK 84], but the authors limited their investigation to objects varying due to vibrations for which the speckle irradiance was naturally varying.

Modern pyroelectric FPAs are made up of ferroelectric pixel arrays joined by bump bonding to similar arrays of silicon, which are used as the readout integrated circuits (ROICs) [KRU 01]. They are modified by an additional radiation chopper which modulates the irradiance in time to allow imaging scenes with stable temperatures. State-of-the-art pyroelectric cameras, such as Pyrocam by Ophir, typically have 320 × 320 pixels with 75 µm pitch and are sensitive from 1 to 3,000 µm wavelengths. Pyroelectric cameras are currently used as IR laser beam profilers. A similar camera, although with a smaller resolution of 124 × 124 pixels, was used for the very first demonstration of LWIR lensless DH [ALL 03].

The other main, and nowadays more prominent, uncooled technology is the one of microbolometer FPAs. A bolometer exhibits electrical resistance variation when its temperature changes due to absorption of light energy [KRU 01]. Microbolometer-based FPAs are made of silicon which is micromachined to obtain a pixel structure. A thin membrane of the pixel size which is made of silicon nitride is attached to the ROIC by means of legs. The detecting material is deposited on the membrane. We distinguish two main detecting materials: vanadium oxide (VO_x) and amorphous silicon (a-Si) [TIS 04]. State-of-the-art microbolometer FPAs can be found at Ulis with 1,024 × 768 pixels with 17 µm pitch. The company Jenoptik proposes a camera with a scanned mask and the Ulis 1,024 × 768 pixels FPA for increasing the resolution to 2,048 × 1,536 pixels, at the expense of the frame rate. The very first use of a microbolometer FPA in holography was demonstrated in 2007 in an in-plane electronic speckle pattern interferometry (ESPI) setup with a 320 × 240 pixels camera [VAN 07] and in lensless DH in 2008 with similar resolution [GEO 08]. The rapidly growing market of uncooled thermal camera with increasing performances at affordable prices (compared to cooled cameras) makes this technology the most attractive for LWIR DH. These cameras are used in most of the applications that will be presented in further sections.

6.3.1.3. *Lenses–windows*

Properties can be found in catalogs or Websites of major manufacturers such as II-VI Infrared. We find two main transparent materials. The first transparent material is germanium which is widely used for LWIR lenses and windows. Useful transmission range of germanium windows is from 2 to 14 μm. It is opaque in the visible. Germanium has the property of thermal runaway, meaning that the transmission decreases as temperature increases. As such, germanium window should be used at temperatures below 100°C. Its index of refraction is about 4 at 10 μm, which makes it of particular interest for high aperture optics. But it requires antireflection coatings if high transmission is required. Nowadays, nearly all objective lenses for LWIR thermographic cameras are made of germanium with antireflection coatings. The second transparent material is zinc selenide (ZnSe) which has a very low absorption coefficient in the range of 0.58–22 μm, which avoids the thermal runaway problem of germanium. ZnSe is often used for lenses and windows because of its low absorption at IR wavelengths as well at some visible ones. Its refractive index is close to 2.4. Thus, it requires antireflection coatings if high transmission is required.

6.3.1.4. *Beamsplitters–beam combiners*

These are important components when developing holographic/interferometric setup. LWIR beamsplitters are generally made of a ZnSe substrate with coating but we can find items with germanium substrate. In most of the cases, the coating incorporates layer(s) of germanium, which make them optically opaque, and then not suitable for combining LWIR beams with visible laser diode beam for alignments. LWIR beamsplitters are typically designed for use at 45° angle of incidence and 10.6 μm wavelength. However, other lines can be achieved. Usually, LWIR beamsplitters are custom made. At 45° incidence, there can be significant differences in the transmittance/reflectance values for *s*- and *p*-polarizations. Most beamsplitters are supplied with a wedge to significantly reduce the interference caused by reflections of the second surface. There is the possibility of polarization insensitive beamsplitters, and some companies have developed specific coatings to allow this feature. LWIR beam combiners often refer to elements which allow combining a visible (red) laser in the IR setup for alignment purposes, in which case ZnSe is the substrate used. The major manufacturers of beamsplitters and beam combiners are II-VI Infrared and ULO Optics.

6.3.1.5. *Polarization components*

Polarization components in LWIR have been studied since the 1960s. One of the first major techniques for producing good polarizers in a broad IR range (from SWIR to LWIR and well above) was to deposit grids of microscopic gold wires onto plastic substrates, allowing polarizers to be very compact, easily rotatable and usable with convergent beams [BIR 60]. The limiting factor is the absorption of plastic substrate, but despite this, they can be found commercially. Dichroic polarizers were also developed on the basis of birefringent calcite properly cut. They were more efficient in the MWIR but were usable around some wavelengths in the LWIR region [BRI 65]. Some other works showed the use of interference effects in thin films (Fabry–Perot effect) to produce high polarization qualities: on polyethylene films (kitchen wrap) where films are stretched for optimizing performance at 10.6 μm [RAM 76], on polycrystalline diamond [CHA 96]. For higher polarization qualities, Brewster angle polarizers have been developed, using various substrates and based on the fact that one of the polarizations is not transmitted for an incidence angle which depends on the substrate refractive index. Such polarizers are not always practical since they induce a lateral shift. Stacks of Brewster angle plates can be used for increasing the polarization quality. Some arrangements in chevron or K-geometries of the Brewster angle window make it possible to obtain no lateral shift of the beam, providing an accurate positioning of the windows [DUM 98]. Traditionally, the commercial polarizers sold by major manufacturers work on the principle of the Brewster angle, with improved properties by the use of thin-film deposition. Polarizers of chevron-type are also made available by the II-VI Infrared company, but they are mainly limited to utilization with small collimated laser beams. Waveplates can also be of great importance for holographic setup. Quarter waveplates and half waveplates can be easily found on the market.

6.3.2. *Specific features of the LWIR domain*

6.3.2.1. *Decreased sensitivity of the holographic setup with respect to displacement*

The most obvious advantage of applying holography in LWIR instead of visible is a decrease in sensitivity during the hologram recording, and a decreased sensitivity to displacements measured in a holographic interferometry experiment. These are reviewed hereafter.

First, let us discuss the stability requirements of holography. One major constraint of holographic, or generally interferometric setup, is that the interference pattern coming from the superposition between the object and reference beams [1.6–1.7] has to be stable during recording by the camera sensor. Any movement in the setup will generate shift of the fringe pattern and the camera sensor will simply record a blurred pattern. As a rule of good practice, we generally consider that the variations of any optical distance in the interferometric setup cannot change more than a fraction of the laser wavelength during recording. We already briefly discussed this in the applicability of cooled IR FPAs, with a cooling system which introduces vibrations in such a way that it prevents correct hologram recording. Considering uncooled camera technologies, this problem does not exist anymore. It is clear that using CO_2 lasers allows holographic setup to be 20 times less sensitive to external perturbations compared to visible setup. This advantage allows building holographic instruments which can be used out of laboratory conditions, e.g. for performing non-destructive testing in hangar conditions [VAN 13] or making holograms of persons situated at long distance [LOC 13].

Second, in holographic interferometry we observe the phase difference $\Delta\varphi = \psi_{r1} - \psi_{r2}$ ([1.69]) between two object states which are recorded as two separate holograms, at instants t_1 and t_2, as is shown in Figure 1.23. The phase difference is observed in an interferogram as is shown in Figure 1.23. The distance between two consecutive fringes depends on the wavelength and the geometry of the setup, through the sensitivity vector $\boldsymbol{S}$ (equation [1.67]). From equation [1.68], we see that a given phase difference $\Delta\varphi$ is proportional to the ratio $(|\boldsymbol{D}|/\lambda)$ between the displacement $|\boldsymbol{D}|$ and the wavelength, considering that the sensitivity vector $\boldsymbol{S}$ has not moved between the two hologram captures. Consequently, if we wish to observe large displacements, a long wavelength is more appropriate than a short wavelength. This is illustrated in Figure 6.1(a) where a phase map modulo 2π has been obtained in LWIR at 9.3 µm. The computed phase was unwrapped to obtain the displacement $|\boldsymbol{D}|$ which is reinjected back into [1.68] with two different wavelengths λ_1 and λ_2. The result is then "rewrapped" to obtain a modulo 2π results that would be obtained if wavelengths λ_1 or λ_2 were used instead of 9.3 µm. Figure 6.1(b) shows the same deformation at λ_1 = 1.85 µm and Figure 6.1(c) at λ_2 = 460 nm. It can be clearly seen that the large deformation cannot be observed in visible wavelengths, whereas it is well adapted to LWIR. This has generated various works in which large displacements were to be measured [VAN 07, VAN 10, ALE 13, GEO 13a, VAN 13].

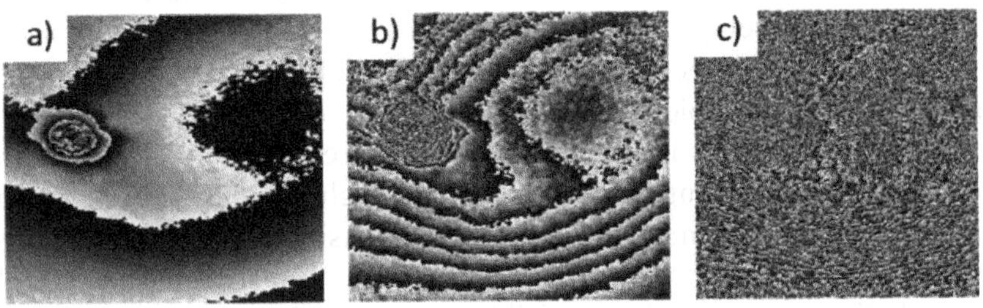

Figure 6.1. *Phase difference mod(2π) of a deformation at various wavelengths a) 9.3 µm, b) 1.85 µm and c) 460 nm*

Besides these two advantages discussed above, LWIR offers other specific advantages or constraints which are developed in the next sections.

6.3.2.2. *Size of reconstructed objects*

In section 1.1.2, the spatial sampling conditions were discussed. The maximum acceptable angle θ_{max} for DH setup, as expressed by [1.8], shows that it is directly governed by the ratio between the wavelength λ and the pixel size $p = p_x = p_y$ (assuming square pixels as is most often the case with LWIR FPAs). When considering the pixel pitch of state-of-the-art FPAs in LWIR and visible, and LWIR wavelengths typically 20 times larger than that of the visible, we have the following formula:

$$\left(\frac{\lambda}{p}\right)_{LWIR} > \left(\frac{\lambda}{p}\right)_{VIS} \quad [6.1]$$

This has the direct effect that, for a given distance of reconstruction, larger objects can be reconstructed in LWIR than in visible, with factors typically ranging between 5 and 10 [PAT 10, GEL 11, VAN 10, GEO 13a]. This has been exploited advantageously for capturing images of objects with human size [GEL 11] or even persons [LOC 13].

In the case of image-plane DH (section 1.4.5), where a lens is used for imaging the object, the observable object size is governed by classical imaging relationship, taking into account the focal length of the imaging lens and its distance to the object. However, the ratio (λ/p) is also important since it governs the useful aperture of the imaging lens through [1.61]. From that equation, we find numerical apertures $\sin\alpha' \leq 0.1$, considering 10 μm wavelength and typical pixel pitches between 15 and 25 μm. In LWIR image-plane DH experiments, which will be presented later, imaging lenses have typical numerical aperture of 1 or 2 (e.g. Tyto or Varia lenses made by Janos) and these numerical apertures cannot be changed. Consequently, the different diffraction orders will overlap and the phase-shifting has to be used to separate them.

6.3.2.3. *Thermal background*

The microbolometer arrays which are used in conjunction with CO_2 lasers in our experiments have a spectral sensitivity ranging from

8 to 14 µm. As is known in fundamental physics, black bodies at a given temperature emit radiation in a certain wavelength range (Planck's law) and Wien's law states that the product of wavelength of maximum spectral radiance and the temperature is a constant equal to 2.8977×10^6 nm. K. It can be calculated that the wavelength of CO_2 lasers corresponds to ambient temperatures around 20°C. Thermal IR radiations emitted by anything in the working environment could on reach the sensor directly or by reflection, either on the object itself or some parts of the setup. Therefore, in LWIR DH, the total irradiance I_{IR} of the pattern recorded by the IR FPAs contains two components. One of them results from the interference between the reference and object beams, i.e. the digital hologram H expressed by [1.6] and [1.7]. The other component I_{Th} contains all the above-mentioned thermal radiations. It is incoherent by nature and constitutes a DC signal. I_{IR} can be expressed as follows:

$$I_{IR} = I_{Th} + H = I_{Th} + a_r^2 + a_o^2 + 2a_r a_o \cos(\varphi_r - \varphi_o) \qquad [6.2]$$

The relatively large number of discretization levels of thermal cameras (typically, 14 or 16 bits) allows us to easily filter out this DC background and keep the useful hologram signal, even if I_{Th} is high with respect to H. Figure 6.2(a) shows the thermal image I_{Th} recorded in an LWIR lensless Fresnel DH setup, with the laser beam switched off ($H = 0$ in [6.2]).

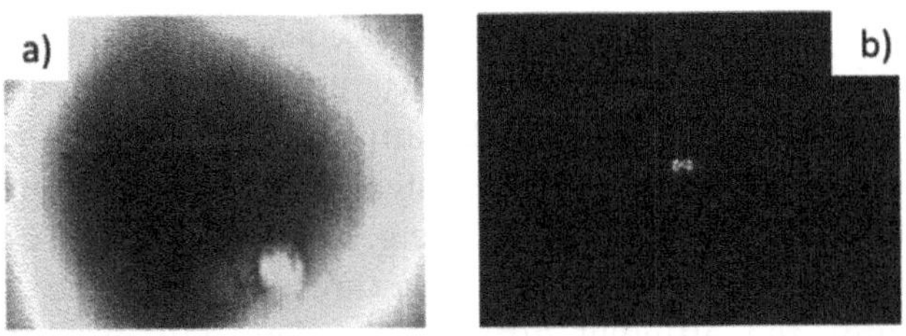

Figure 6.2. *a) Typical thermal background in a lensless Fresnel DH setup and b) its contribution to the numerical reconstruction. For a color version of this figure, see www.iste.co.uk/picart/digiholography.zip*

When injecting I_{Th} in the discrete Fresnel transform and using a typical reconstruction distance d of 1 m, we obtained a central peak shown in Figure 6.2(b). The extent of this peak will strongly depend on the presence of any object acting as a heat source in the field of view, and therefore we cannot draw any conclusions from the example shown. Therefore, we propose to filter this noise in the same way as other authors remove noise in DH [KRE 97], in particular we consider removal of halo, reference and object (HRO) components as discussed by Skotheim [SKO 03]. The HRO subtraction consists of using modified hologram intensity H' in place of H in the discrete Fresnel transform and from which images of reference and object beams, respectively, a_r^2 and a_o^2, are deduced. These images are captured independently by using shutters in both beams, and prior to hologram capture. We can use the same approach but also removing the thermal background, applying so a "HROT subtraction" [GEO 13a]. The intensity of the hologram to be injected into the numerical reconstruction is then given by:

$$H' = H - a_r^2 - a_o^2 - I_{Th} \qquad [6.3]$$

However, the latter is not exactly true because the thermal background I_{Th} also affects the reference and object beams. Assuming that I_{Th} has not changed between acquisitions, we can deduce it from the measurements of other two beams. If we add the suffix m to designate the measured images, we can write the hologram intensity to be used in the reconstruction as follows:

$$\begin{aligned} H' &= H_m - \left[a_{r,m}^2 - I_{Th,m}\right] - \left[a_{o,m}^2 - I_{Th,m}\right] - I_{Th,m} \\ &= H_m - a_{r,m}^2 - a_{o,m}^2 + I_{Th,m} \end{aligned} \qquad [6.4]$$

This thermal background can also be considered as an interesting signal and not necessarily a noise to be removed. This will be presented later in the chapter.

6.3.2.4. *Type of reflectivity of objects*

The LWIR wavelengths are typically 20 times larger than visible ones. Therefore, an object which is scattering when observed in visible can be more (if not totally) specular in LWIR. Yamaguchi [YAM 03] has shown that when the wavelength is close to the surface roughness or smaller, the surface becomes completely scattering and speckle appears. On the other side, specular reflection appears as soon as the wavelength is larger than the roughness and quickly dominates the speckle pattern intensity. Nevertheless, up to some extent, the speckle pattern can still be observed but its intensity is smaller than the specular peak. The total integrated scatter (TIS) represents the scattered light intensity normalized by the intensity of the reflected beam. It is proportional to the square of the ratio between the roughness and the wavelength [STO 90]. Consequently, if the wavelength is 20 times larger, the TIS will be 400 times smaller. Thus, the speckle intensity will be 400 times smaller using a CO_2 laser than with an equivalent experiment using a visible laser.

For surfaces with roughness in the range of a few nanometers (e.g. metals), we can deal with specularity by spraying the surface with a white removable powder (developer for liquid dye penetrant non-destructive technique) [VAN 10] which is largely used by holographers in the visible range. Some materials have a natural roughness which is of the same order than the CO_2 laser wavelengths [VAN 09]. This allows for producing speckles in LWIR without spraying the surface and makes easily feasible the application of LWIR holographic methods [ALE 13]. In the intermediary cases, where both specular and scattering reflectivities are observed, we have to be highly cautious to avoid specular peaks that could endanger the microbolometer array. Damage threshold of microbolometer and pyroelectric arrays were measured in view of the use in IR laser beam profiling [PIC 01]. However, in holographic experiments, like those presented here, it is difficult to evaluate the irradiance of spurious spots coming in the thermal camera by reflections on the object or any other part of the setup. Therefore, we have to be extremely cautious when setting up an LWIR holographic experiment.

The situation is depicted in Figure 6.3(c) where a diverging beam illuminates an object and the reflected specular peak avoids entering into the imaging system. The irradiance (in arbitrary unit) of the object image on the FPA is shown in Figure 6.3(a) on a pixel line for various laser powers. The specular peak is easily visible. The thermal background is seen for zero power of laser, and Figure 6.3(b) shows a vertical zoom of Figure 6.3(a) where the speckle besides the specular peak is easily observed for various laser powers, on the right-hand side of pixel #51. The procedure we adopt for not endangering the sensor is to work at low power and progressively increase the laser power. When specular peak begins to appear, like in Figure 6.3(a), the object is tilted to place the peak out of the sensor. Often the peak coming from the object is prominent. However, some other parts of the setup (e.g. mechanical mounts) can provide back-reflections which can be dangerous, even on very limited pixel area. For helping in alignments, we use a cheap handheld low-resolution IR camera and movable screens in paper which allows seeing where back-reflections arrive in the setup.

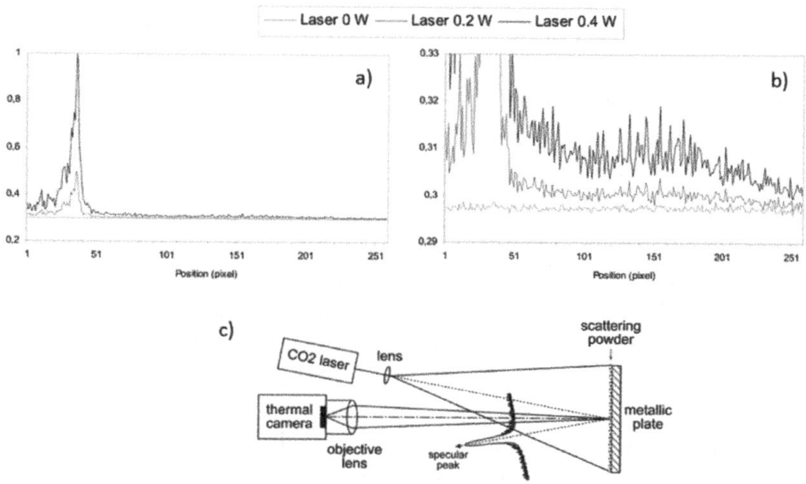

Figure 6.3. *a) Irradiance (in arbitrary unit) of specular peak received by a microbolometer array at various laser powers, b) zoom of a showing the speckle out of the specular peak and c) oblique incidence of illumination to avoid specular peak in the camera*

6.4. Typical applications of LWIR digital holography

6.4.1. *Recording holograms of large objects in LWIR and display in visible*

Displaying live three-dimensional (3D) images, in real time, of an object situated in a remote location is possible. For that, a digital hologram is recorded and stored on computer and, if a spatial light modulator (SLM) is addressed by this pattern, a laser beam which illuminates it will diffract a beam containing the image information [SUT 00, MAT 02, KOH 06]. This principle has been applied largely, making use of recordings and read-outs at visible wavelengths. Some have demonstrated that the digital hologram, once recorded, can be sent via the Internet to be reconstructed in another country. The interest of using LWIR for recording the hologram of large objects has been explained in various papers. The holograms of statues a few tens of centimeters high [PAT 10, PEL 10a, PEL 10b], or mannequins up to the human size, have been shown [GEL 11]. The setup is shown in Figure 6.4 where the upper part shows the hologram recording with the CO_2 laser in the Fourier configuration (for limiting aberrations) as is noted in [PEL 10a]. In the reconstruction process, they use a diode-pumped solid state (DPSS) laser emitting at 532 nm which is expanded and made converging before reaching the SLM. An image is then produced which is observed through a beamsplitter (BS) and a folding mirror (M3) by either a charge-coupled device (CCD) camera or directly on a screen. The laser power used in the experiment is 30 W, the camera is based on an uncooled microbolometer a-Si array, with 640 × 480 pixels of 25 μm pitch, model Miricle made by Thermoteknix. The SLM is based on liquid crystal on silicon (LCOS), model Pluto by Holoeye, with 1,920 × 1,080 pixels [PAT 10].

The authors also note the advantage of DH with respect to the huge difference between the wavelengths used for recording and readout. Indeed, changing the wavelength between both steps leads to aberrations and distortions in the diffracted image; the latter depends on the wavelength ratio and difference in pixel pitches. In DH, this is no longer an issue since the authors apply numerical transformation such as stretching and scale changes to overcome this [PEL 10a]. The possibility of digital holograms stitching has been proposed and it allows us to further increase the size of reconstructed objects up to human size [PEL 12]. Figure 6.5(a) shows the LWIR part of the setup used for recording the hologram of a 33 cm high statuette seen in the forefront on the left. Figure 6.5(b) shows the image reconstructed by the SLM with the 532 nm DPSS laser and which is projected on a screen. Applications of this technique are foreseen for virtual museums for which circular displays combining several SLMs have been demonstrated [PAT 11, STO 12].

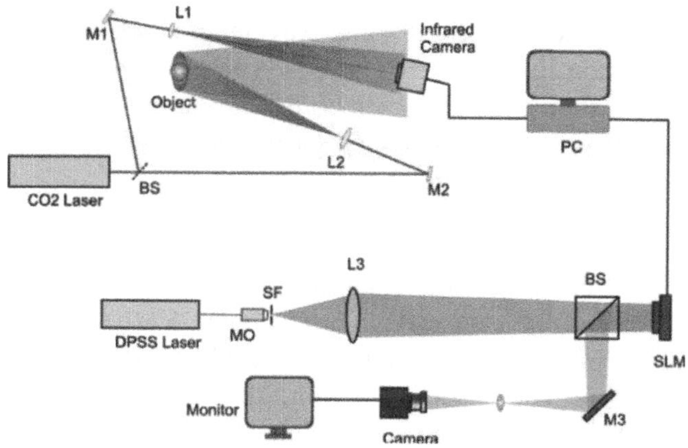

Figure 6.4. *Setup for recording large object hologram with LWIR Fourier DH and display in visible (courtesy of Ferraro) [PEL 10a]. For a color version of this figure, see www.iste.co.uk/picart/digiholography.zip*

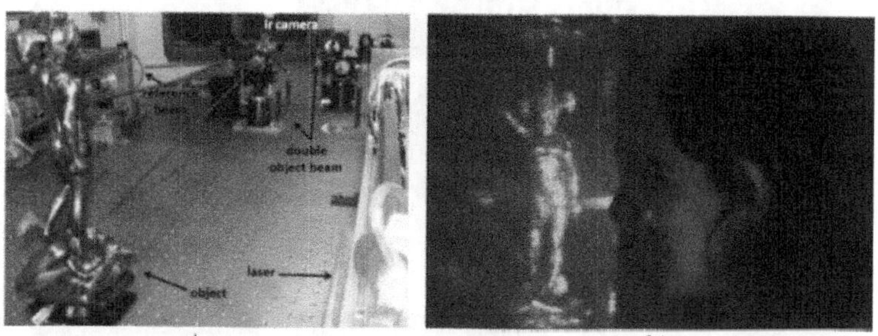

a) b)

Figure 6.5. *a) Picture of the LWIR part of the setup used for recording the statuette hologram and b) display of the live image using the SLM and DPSS laser (courtesy of Ferraro). For a color version of this figure, see www.iste.co.uk/picart/digiholography.zip*

6.4.2. *Reconstruction of images through smoke and flames*

An original application of LWIR DH has been proposed by Locatelli *et al.* [LOC 13]. Since lensless DH can be used to perform out-of-focus acquisitions and the numerical reconstruction allows us to reconstruct scenes at any distance, we should be able to reconstruct a scene or an object which is located behind or inside smoke, provided that it is transparent to the wavelength used. Locatelli *et al.* note that CO_2 and H_2O gases absorb at some wavelengths which are different from the emission lines of the CO_2 laser. Therefore, a DH setup based on such lasers should be able to record a hologram of objects which are not seen in visible through such gases and which can be found in some industrial environments or in fire scenes.

Figure 6.6(a) shows the part of the set which was built in a laboratory for creating smoke around the object. The object is a statuette enclosed in a plexiglass box. Figure 6.6(b) shows the box filled with smoke. The statuette is invisible to the human eye; however, it can be observed with the thermographic camera (Figure 6.6(c)). The laser and setup components are the same as those presented in the previous section. Figure 6.6(d) shows the image reconstructed numerically in the plane where the object stands in the box.

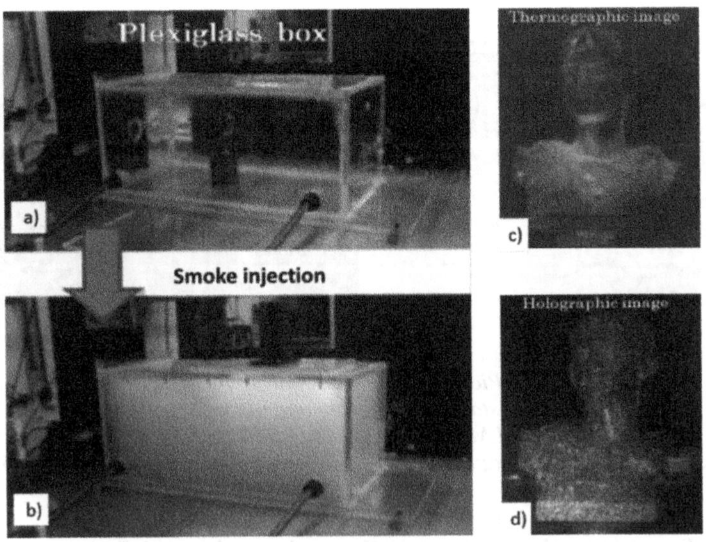

Figure 6.6. *a) Statuette in plexiglass box, b) box filled with smoke, c) thermographic image of the statuette through smoke and d) image reconstructed by DH (courtesy of Ferraro). For a color version of this figure, see www.iste.co.uk/picart/digiholography.zip*

In the second experiment, the authors place the same object behind the flames of some candles. In classical thermography, as is shown in Figure 6.7(a), the flames saturate the camera sensor, preventing correct observation of the object, the object having generally a lower temperature than the flames. The alternative setup with LWIR lensless DH is shown in Figure 6.7(b). The flame does not provide an image on the detector. Instead, light rays emitted by the laser will be scattered by the object and some of them will form the hologram. Incoherent light from the flame does not participate in the hologram. This allows for reconstructing the object located behind the heat source without saturating the detector (Figure 6.7(d)).

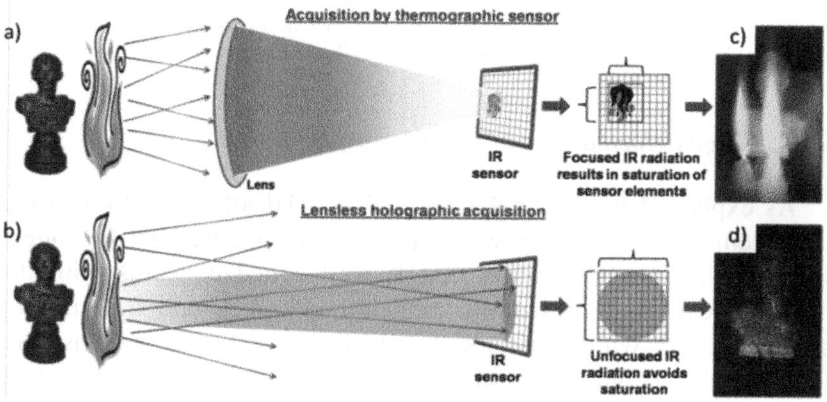

Figure 6.7. *Comparison between a) classical thermographic imaging and b) LWIR lensless DH setup. c) Object blinded by flame in classical thermography and d) image reconstructed by LWIR DH (courtesy of Ferraro). For a color version of this figure, see www.iste.co.uk/picart/digiholography.zip*

Figure 6.8(a)–(c) shows results obtained with living persons located behind flames which are not visible by direct thermographic imaging. Figure 6.8(d) shows the images reconstructed by LWIR lensless DH.

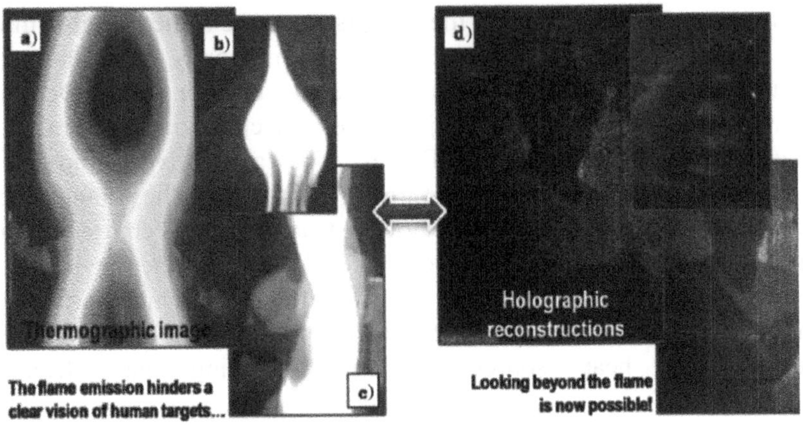

Figure 6.8. *a)–c) Various images obtained by classical thermography and blinded by the head of flames and d) images reconstructed by LWIR DH (courtesy of Ferraro). For a color version of this figure, see www.iste.co.uk/picart/digiholography.zip*

6.4.3. *Large deformations of specular aspheric reflectors*

As explained in section 6.3.2.1, LWIR DHI allows us to measure displacements larger than with visible DHI because of the longer wavelength. One application that is well suited concerns with the measurement of deformations ranged between 1 and 250 μm of space reflectors with diameters of up to 4 m. These deformations have to be measured during thermal-vacuum balance tests in large space simulators. A trade-off was made between LWIR DHI and other techniques, like LWIR classical interferometry or methods based on imagery and triangulation [GEO 13a]. The advantage of using DHI is that it allows following the object deformation in time without considering an external reference like in interferometry, which is an advantage for aspheric surfaces, which otherwise would require expensive null lenses. Using LWIR DHI will provide a desensitized measurement compared to visible DHI. However, obtaining such resolution in displacements using methods based on imagery (such as fringe projection, digital image correlation and videogrammetry) would be difficult. Indeed, the resolution of the latter does not depend on the wavelength but on the number of pixels and the volume observed. We concluded from the trade-off that LWIR DHI fills a gap between visible holography and imagery-based techniques for such large objects [GEO 13a]. A characteristic of the specimens to be measured is that they are already specular in the visible, hence in LWIR. Such space optics cannot be sprayed by scattering removable coating because post-cleaning could damage their delicate surfaces. Therefore, we had to cope with this in the experimental developments reported hereafter.

We developed a Fresnel DH setup with a 10 W CO_2 laser and a Variocam HR camera from Jenoptik, with 640 × 480 pixels. The interferometer had to be placed under the vacuum chamber, with the reflector under test. Since the laser and the camera are not vacuum compatible, we had to place them outside the chamber, while the remaining part of the interferometer is inside. Also, the highly specular behavior of these surfaces led us to study the different ways in which to illuminate the reflector to form an object beam which could interfere with the reference beam on the FPA. Figure 6.9 shows a diagram of the LWIR DH setup. In the case of an astigmatic reflector, like parabola, using a point source will give an extended image spot least confusion circle (LCC) which can interfere with a reference beam, as shown in Figure 6.9(a). The reference beam is formed by the beamsplitter BS1 and travels through a mirror MPZT and is reflected by the beam combiner BC to the camera. The LCC is imaged on the camera sensor S through a pair of lenses. This configuration is referred to as the specular configuration because a given ray impinging the object gives rise to a well-identified ray on the FPA. We can also illuminate with an extended source formed by a reflective diffuser D, as shown in Figure 6.9(b). This configuration is referred to as the scattering configuration. An extended spot is also created and is imaged on S, similarly to the other case. Both configurations were described and compared in detail in [GEO 13a, VAN 14a]. It was concluded that the point source illumination gives a better quality interferogram than the extended source. However, the latter does not suffer from vignetting in the case of misalignment and it allows for a larger displacement measurement range. The versatility of the extended source illumination was successfully used for testing other aspherics, such as ellipses [GEO 13b, VAN 14b].

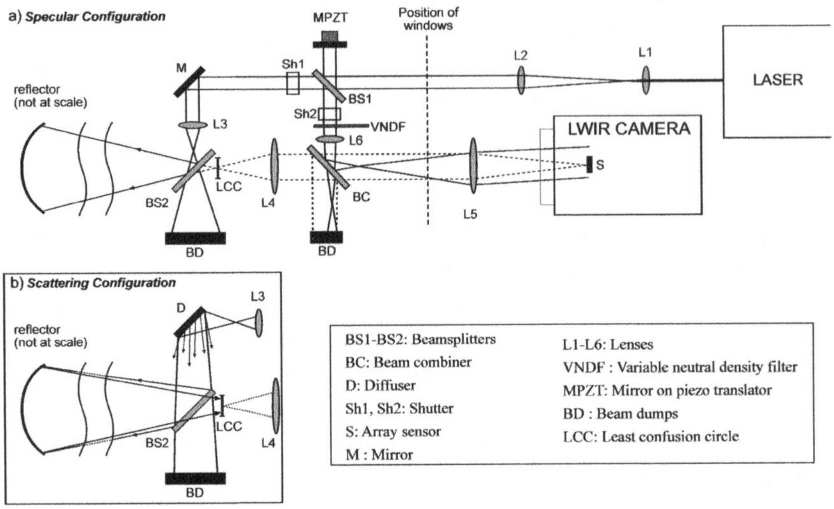

Figure 6.9. *Diagram of the LWIR DHI setup for aspheric testing in vacuum chamber*

An in-line diagram was used instead of an off-axis diagram because it allows for increasing the spatial resolution. The phase-shifting reconstruction principle was needed to be used to obtain the +1 order of interest (see section 1.1.3.2); it was made possible due to the piezotranslator mounted on mirror MPZT. Figure 6.10 shows the implementation of the above setup in the vacuum chamber and some results obtained. Figure 6.10(a) shows the 5 m diameter facility used for the test at the *Centre Spatial de Liège* (Belgium), and Figure 6.10(b) shows the parabolic reflector used as specimen, and surrounded by a thermal shroud for inducing temperature changes. Figure 6.10(c) shows the diagram of the setup, placed in the chamber, except for some components which were kept outside (mainly, laser and camera). Figure 6.10(d) shows a typical phase difference modulo 2π obtained by subtracting phases reconstructed by DH and corresponding to object states at 107 and 295 K. The resulting deformation is shown in Figure 6.10(e).

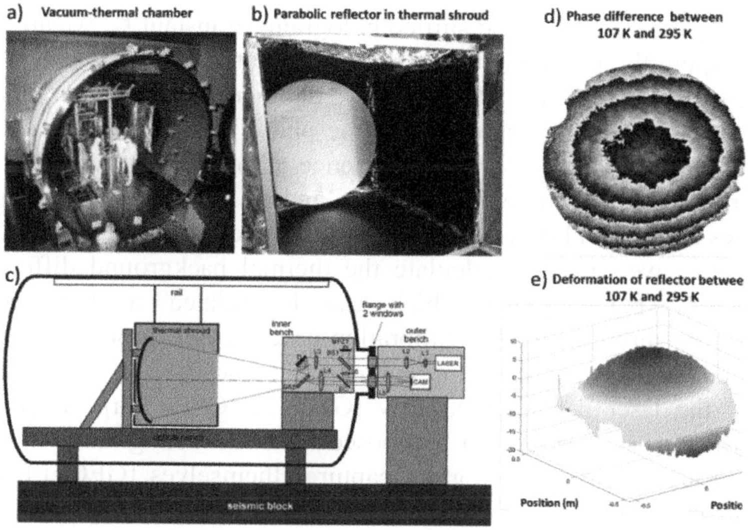

Figure 6.10. *a) Vacuum chamber, b) parabolic reflector in thermal shroud, c) diagram of the vacuum chamber (not at scale) with the setup, d) phase difference due to thermal change and e) final deformation. For a color version of this figure, see www.iste.co.uk/picart/digiholography.zip*

6.4.4. *Combined holography and thermography for thermomechanical analysis and non-destructive testing*

As discussed in section 6.3.2.3, LWIR digital holograms are affected by a thermal background I_{Th} which can be eliminated if needed. However, it can be interesting to use I_{Th} if we wish to measure the temperature of the object at the instant of the hologram capture. This can be of particular interest in numerous engineering applications, such as thermomechanical analyses where deformations are related to temperature differences. As the thermal background is incoherent, it cannot be reconstructed by any of the DH principles. Only an image-plane configuration can be used for creating a true thermal image in addition to the hologram. We have presented different steps of setup development in view of this application. Off-axis LWIR Fresnel DHI was studied first [ALE 10] and then LWIR ESPI, which is basically an in-line image-plane DH configuration associated with phase-shifting [GEO 12, ALE 13].

If we record four phase-shifted holograms at instant t_1, we are able to compute the phase related to the object state. When the object has undergone some stress at instant t_2, another series of four phase-shifted holograms is captured and the phase is calculated again. As usual, calculating the phase difference allows us to observe the deformation of the object between t_1 and t_2. If thermal background images $I_{Th}(t_1)$ and $I_{Th}(t_2)$ are captured just before each phase-shifted sequence, we can also calculate the thermal background difference $\Delta I_{Th} = I_{Th}(t_2) - I_{Th}(t_1)$ which can be related to temperature difference through suitable calibration process. For these preliminary thermal images recordings, a shutter has to be used in the setup to block the laser beam temporarily [GEO 12, GEO 14a]. A specific algorithm has also been proposed for extracting the thermal background from the hologram captures themselves [GEO 14b]. In any case, we have a combination of holographic deformation and temperature variations obtained simultaneously (or quasi-simultaneously) and in every pixel. This has the advantage of not requiring two separate measurement systems which would require correlation or data fusion between two sets of results. Figure 6.11(a) shows an example where a helicopter panel with a repair is heated by two halogen lamps. Figure 6.11(b) shows the phase difference (modulo 2π), Figure 6.11(c) shows a gray level picture of the corresponding deformation and Figure 6.11(d) shows a 3D plot. The corresponding temperature variation is shown in Figure 6.11(e). Figure 6.11(f) shows a hybrid representation of the 3D view of the deformation colored with respect to the temperature variation.

In order to use such techniques in field conditions, we showed the achievement of a mobile system which incorporates the laser, the camera (Jenoptik Variocam HR) and all necessary components [VAN 13]. Figure 6.12 shows the diagram of the instrument. It is composed of two parts: an upper bench (Figure 6.12(a)) with the camera and the optical elements constituting the reference beam. The latter comes from the polarizing beamsplitter PBS and is expanded by lens L and folded by MPZT toward a concave mirror CM which reflects the beam through the lens OL attached to the camera. A thorough study of constraints about the reference beam [ALE 13, VAN 13] led us to consider such an arrangement. The lower bench

(Figure 6.12(b)) contains the compact CO_2 laser (model Merit-S from Access Laser Co) and the PBS. The latter, made by II-VI Infrared, consists of a stack of ZnSe plates with a chevron geometry and is shown in Figure 6.12(c). The final system is shown in Figure 6.13(a) in front of a large composite structure lying in the testing facility of a major aeronautical manufacturer.

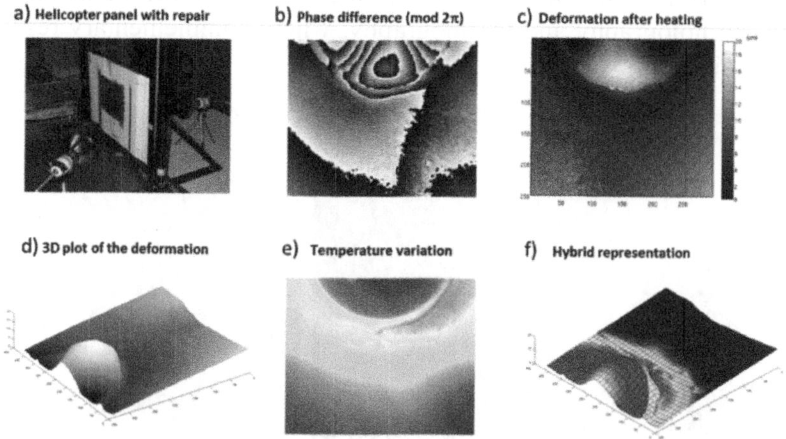

Figure 6.11. *Example of separation of deformation and temperature variations. For a color version of this figure, see www.iste.co.uk/picart/digiholography.zip*

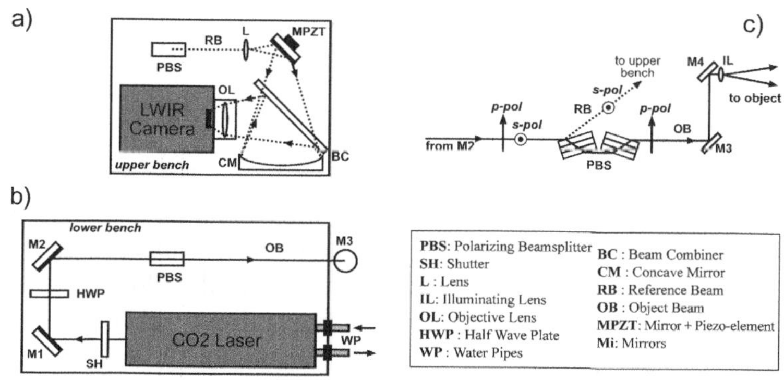

Figure 6.12. *Diagram of the mobile LWIR ESPI system: a) upper bench with camera and combiner, b) lower bench with laser and beamsplitter and c) detail of the beam separation*

Figure 6.13(b) and (c) show an example of non-destructive testing applications where an observed area of the large composite structure has been heated with halogen lamps. A subsurface defect appears after 30 s of heating in the deformation, encircled in red (Figure 6.13(b)). Other subsurface defects appear in the thermal variation image (Figure 6.13(c)) obtained at the same time; they are encircled in yellow. This shows a really interesting feature of the combination between holography and thermography, where complementary results are obtained.

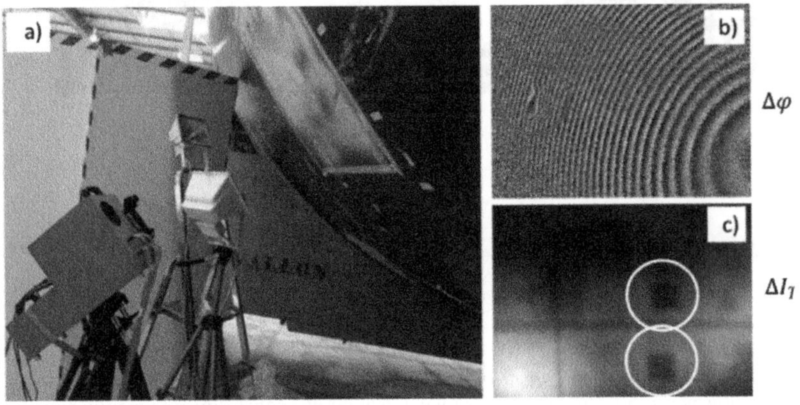

Figure 6.13. *a) Mobile LWIR ESPI system working in industrial testing facility. b) Phase difference showing the deformation due to heating and c) corresponding temperature difference. Different subsurface defects are observed in each component. For a color version of this figure, see www.iste.co.uk/picart/digiholography.zip*

6.5. Conclusions: future prospects

In this chapter, we have shown some interesting outcomes of using LWIR lasers, emitting around 10 µm, and thermal cameras sensitive in that range, in different configurations of digital holography. After reviewing the specific components used at these wavelengths, we discussed some specific aspects and constraints of the considered spectral range. These specificities have been used for demonstrating innovative applications of digital holography, which could not be possible if shorter wavelengths were used. Among them, let us cite the possibility of capturing holograms of living people hidden by smoke and flames in fire incidents and the combination of holography and thermography in a single sensor for improved non-destructive testing. Also, the long wavelength offers the double advantage that it can be used for measuring large displacements or deformations and it allows better immunity of holography against external perturbations. A mobile LWIR holographic system has been demonstrated to work in industrial plant environments.

In the future, the domain of LWIR DH will certainly open itself to a broader range of applications. Let us cite the recent demonstration of vibration mode shapes observation [VAN 14c] and the use of tunable quantum cascade lasers which was shown in two-wavelength DHI [RAV 14]. The constant evolution of critical components, such as lasers and cameras, will certainly push further, in particular to make more compact systems, together with an affordable price.

6.6. Bibliography

[ALE 10] ALEXEENKO I., VANDENRIJT J.-F., GEORGES M.P., *et al.*, "Digital holographic interferometry by using long wave infrared radiation (CO_2 laser)", *Applied Mechanics and Materials*, vol. 24–25, pp. 147–152, 2010.

[ALE 13] ALEXEENKO I., VANDENRIJT J.-F., PEDRINI G., *et al.*, "Nondestructive testing by using long-wave infrared interferometric techniques with CO_2 lasers and microbolometer arrays", *Applied Optics,* vol. 52, no. 1, pp. A56–A67, 2013.

[ALL 03] ALLARIA E., BRUGIONI S., DE NICOLA S., *et al.*, "Digital holography at 10.6 μm", *Optics Communications*, vol. 215, pp. 257–262, 2003.

[ARM 14] ARMSTRONG J.M., SKOKAN M.R., KINCH M.A., *et al.*, "HDVIP five-micron pitch HgCdTe focal plane arrays", *Proceedings of the International Society for Optics and Photonics (SPIE)*, vol. 9070, p. 907033, 2014.

[ASW 03a] ASWENDT P., "Micromotion analysis by deep-UV speckle interferometry", *Proceedings of SPIE*, vol. 5145, pp. 17–22, 2003.

[ASW 03b] ASWENDT P., "Promise and challenge of DUV speckle interferometry", *Proceedings of SPIE*, vol. 4933, pp. 117–122, 2003.

[BEA 77] BEAULIEU R., LESSARD R.A., CORMIER M., *et al.*, "Infrared holography on commercial wax at 10.6 µm", *Applied Physics Letters*, vol. 31, no. 9, pp. 602–603, 1977.

[BEA 78] BEAULIEU R., LESSARD R.A., CORMIER M., *et al.*, "Pulsed IR holography on takiwax films", *Applied Optics*, vol. 17, no. 22, pp. 3619–3621, 1978.

[BEA 94] BEAULIEU R., LESSARD R.A., LING CHIN S., "Resist recording media for holography at 10.6 µm", *Proceedings of SPIE*, vol. 2042, pp. 259–263, 1994.

[BEA 00] BEAULIEU R., LESSARD R.A., "Infrared holography on poly(acrylic acid) films", *Proceedings of SPIE*, vol. 4087, pp. 1298–1301, 2000.

[BIR 60] BIRD G., PARRISH M., "The wire grid as a near-infrared polarizer", *Journal of Optical Society of America*, vol. 50, no. 9, pp. 886–891, 1960.

[BRI 65] BRIDGES T., KLUVER J., "Dichroic calcite polarizers for the infrared", *Applied Optics*, vol. 4, no. 9, pp. 1121–1125, 1965.

[CAL 03] CALIXTO S., "Albumen as a relief recording media for spatial distributions of infrared radiation. Fabrication of interference gratings and microlenses", *Applied Optics*, vol. 42, no. 2, pp. 259–263, 2003.

[CHA 96] CHARDONNET C., BERNARD V., DAUSSY C., *et al.*, "Polarization properties of thin films of diamond", *Applied Optics*, vol. 35, no. 34, pp. 6692–6697, 1996.

[CHE 87] CHEO P.K., *Handbook of Molecular Lasers*, Optical Engineering Series, Marcel Dekker, New York, vol. 14, 1987.

[CHI 69] CHIVIAN J.S., CLAYTOR R.N., EDEN D.D., "Infrared holography at 10.6 µm", *Applied Physics Letters*, vol. 15, pp. 123–125, 1969.

[CON 76] CONKLIN T., STUPP E.H., "Applications of the pyroelectric vidicon", *Optical Engineering*, vol. 15, no. 6, pp. 510–515, 1976.

[DEC 72] DECKER G., HEROLD H., RÖHR H., "Holography and holographic interferometry with pulsed high power infrared lasers", *Applied Physics Letters*, vol. 20, no. 12, pp. 490–492, 1972.

[DEL 09] DELACRÉTAZ Y., BERGOËND I., DEPEURSINGE C., "Digital holographic microscopy for micro-systems investigation in near infrared", *3rd EOS Topical Meeting on Optical Microsystems*, Capri, Italy, 2009, available at http://www.digitalholography.eu/varasto/capri_oms09_-_Delacretaz_Yves_-_2302.pdf.

[DEL 10] DELACRÉTAZ Y., DEPEURSINGE C., "Digital holographic microscopy for silicon microsystems metrology", *Proceedings of SPIE*, vol. 7719, p. 77191M, 2010.

[DIA 07] DIAKIDES N., BRONZINO J., *Medical Thermal Imaging*, CRC Press, 2007.

[DUC 11] DUCKWORTH T.A., OGRIN F., DHESI S.S., *et al.*, "Magnetic imaging by x-ray holography using extended references", *Optics Express*, vol. 19, no. 17, pp. 16223–16228, 2011.

[DUM 98] DUMMER D., KAPLAN S., HANSSEN L., *et al.*, "High-quality Brewster's angle polarizer for broadband infrared application", *Applied Optics*, vol. 37, no. 7, pp. 1194–1204, 1998.

[EIS 04] EISEBITT S., LÜNING J., SCHLOTTER W.F., *et al.*, "Lensless imaging of magnetic nanostructures by X-ray spectro-holography", *Nature*, vol. 432, pp. 885–888, 2004.

[FAR 10] FARIDIAN A., HOPP D., PEDRINI G., *et al.*, "Nanoscale imaging using deep ultraviolet digital holographic microscopy", *Optics Express*, vol. 18, no. 13, pp. 14159–14164, 2010.

[FOR 73] FORMAN P.R., HUMPHRIES S., PETERSON R.W., "Pulsed holographic interferometry at 10.6 μm", *Applied Physics Letters*, vol. 22, no. 10, pp. 537–539, 1973.

[GEL 11] GELTRUDE A., LOCATELLI M., POGGI P., *et al.*, "Infrared digital holography for large objects investigation", *Proceedings of SPIE*, vol. 8082, p. 80820C, 2011.

[GEO 08] GEORGE N., KHARE K., CHI W., "Infrared holography using a microbolometer array", *Applied Optics,* vol. 47, no. 54, pp. A7–A12, 2008.

[GEO 12] GEORGES M.P., VANDENRIJT J.-F., THIZY C., *et al.*, "Speckle interferometry at 10 μm with CO_2 lasers and microbolometers array", *Proceedings of SPIE*, vol. 8412, p. 84121O, 2012.

[GEO 13a] GEORGES M.P., VANDENRIJT J.-F., THIZY C., *et al.*, "Digital holographic interferometry with CO_2 lasers and diffuse illumination applied to large space reflector metrology [invited]", *Applied Optics*, vol. 52, no. 1, pp. A102–A116, 2013.

[GEO 13b] GEORGES M.P., VANDENRIJT J.-F., THIZY C., *et al.*, "Digital holographic interferometry with CO_2 laser applied to aspheric space reflector testing", *Digital Holography and Three-Dimensional Imaging Conference*, Hawaii, USA, 2013, available at http://www.opticsinfobase.org/abstract.cfm?URI=DH-2013-DW3A.4.

[GEO 14a] GEORGES M.P., VANDENRIJT J.-F., THIZY C., *et al.*, "Speckle interferometry at 10 micrometers wavelength: a combined thermography and interferometry technique and its application in aeronautical nondestructive testing", *Proceedings of SPIE*, vol. 9204, p. 92040F, 2014.

[GEO 14b] GEORGES M.P., VANDENRIJT J.-F., THIZY C., *et al.*, "Combined holography and thermography in a single sensor through image-plane holography at thermal infrared wavelengths", *Optics Express*, vol. 22, no. 21, pp. 25517–25529, 2014.

[JOR 13] JORGE ALDAVE I., VENEGAS BOSOM P., VEGA GONZÁLEZ L., *et al.*, "Review of thermal imaging systems in composite defect detection", *Infrared Physics & Technology*, vol. 61, pp. 167–175, 2013.

[KOB 71] KOBAYASHI S., KURIHARA K., "Infrared holography with wax and gelatin film", *Applied Physics Letters*, vol. 19, pp. 482–484, 1971.

[KOH 06] KOHLER C., SCHWAB X., OSTEN W., "Optimally tuned spatial light modulators for digital holography", *Applied Optics*, vol. 45, no. 5, pp. 960–967, 2006.

[KRE 97] KREIS T., JUEPTNER W., "Suppression of the DC term in digital holography", *Optical Engineering*, vol. 36, pp. 2357–2360, 1997.

[KRE 05] KREIS T., *Handbook of Holographic Interferometry: Optical and Digital Methods*, Wiley, New York, 2005.

[KRU 01] KRUSE P.W., *Uncooled Thermal Imaging Arrays, Systems and Applications*, Tutorial Text Series, SPIE, Bellingham, vol. TT51, 2001.

[KWO 80] KWON O., WYANT J.C., HAYSLETT C.R., "Rough surface interferometry at 10.6 μm", *Applied Optics*, vol. 19, no. 11, pp. 1862–1869, 1980.

[LEV 93] LEVINE B.F., "Quantum-well infrared photodetectors", *Journal of Applied Physics*, vol. 74, no. 8, pp. R1–R81, 1993.

[LEW 84] LEWANDOWSKI J., MONGEAU B., CORMIER M., "Real time interferometry using IR holography on oil films", *Applied Optics*, vol. 23, no. 2, pp. 242–246, 1984.

[LEW 86] LEWANDOWSKI J., MONGEAU B., CORMIER M., *et al.*, "Infrared holographic interferometry", *Applied Optics*, vol. 25, no. 18, pp. 3291–3296, 1986.

[LOC 13] LOCATELLI M., PUGLIESE E., PATURZO M., *et al.*, "Imaging live humans through smoke and flames using far-infrared digital holography", *Optics Express*, vol. 21, no. 5, pp. 5379–5390, 2013.

[LØK 84] LØKBERG O.J., KWON O., "Electronic speckle pattern interferometry using a CO_2 laser", *Optics & Laser Technology*, vol. 16, no. 4, pp. 187–192, 1984.

[MAT 02] MATOBA O., NAUGHTON T.J., FRAUEL Y., *et al.*, "Real-time three-dimensional object reconstruction by use of a phase-encoded digital hologram", *Applied Optics*, vol. 41, no. 29, pp. 6187–6192, 2002.

[PAT 10] PATURZO M., PELAGOTTI A., FINIZIO A., *et al.*, "Optical reconstruction of digital holograms recorded at 10.6 µm: route for 3D imaging at long infrared wavelengths", *Optics Letters*, vol. 35, pp. 2112–2114, 2010.

[PAT 11] PATURZO M., PELAGOTTI A., GELTRUDE A., *et al.*, "Infrared digital holography applications for virtual museums and diagnostics of cultural heritage", *Proceedings of SPIE*, vol. 8084, p. 80840K, 2011.

[PED 07] PEDRINI G., ZHANG F., OSTEN W., "Digital holographic microscopy in the deep (193 nm) ultraviolet", *Applied Optics*, vol. 46, no. 32, pp. 7829–7835, 2007.

[PEL 10a] PELAGOTTI A., PATURZO M., GELTRUDE A., *et al.*, "Digital holography for 3D imaging and display in the IR range: challenges and opportunities", *3D Research*, vol. 1, no. 4, pp. 1–10, 2010.

[PEL 10b] PELAGOTTI A., LOCATELLI M., GELTRUDE A., *et al.*, "Reliability of 3D imaging by digital holography at long IR wavelength", *Journal of Display Technology*, vol. 6, no. 10, pp. 465–471, 2010.

[PEL 12] PELAGOTTI A., PATURZO M., LOCATELLI M., *et al.*, "An automatic method for assembling a large synthetic aperture digital hologram", *Optics Express*, vol. 20, no. 5, pp. 4830–4839, 2012.

[PIC 01] PICARD F., JEROMINEK H., POPE T.D., *et al.*, "Nonimaging applications for microbolometer arrays", *Proceedings of SPIE*, vol. 4369, pp. 274–286, 2001.

[RAM 76] RAMPTON D., GROW R., "Economic infrared polarizer utilizing interference effects in films of polyethylene kitchen wrap", *Applied Optics*, vol. 15, no. 4, pp. 1034–1036, 1976.

[RAV 14] RAVARO M., LOCATELLI M., PUGLIESE E., *et al.*, "Mid-infrared digital holography and holographic interferometry with a tunable quantum cascade laser", *Optics Letters*, vol. 39, no. 16, pp. 4843–4846, 2014.

[REP 05] REPETTO L., CHITTOFRATI R., PIANO E., *et al.*, "Infrared lensless holographic microscope with a vidicon camera for inspection of metallic evaporations on silicon wafers", *Optics Communications*, vol. 251, pp. 44–50, 2005.

[RIO 77] RIOUX M., BLANCHARD M., CORNIER M., *et al.*, "Plastic recording media for holography at 10.6 μm", *Applied Optics*, vol. 16, no. 7, pp. 1876–1879, 1977.

[ROB 76] ROBERTS R.R., BLACK T.D., "Infrared holograms recorded at 10.6 μm and reconstructed at 0.6328 μm", *Applied Optics*, vol. 15, no. 9, pp. 2018–2019, 1976.

[ROG 05] ROGALSKI A., "HgCgTe infrared detector material: history, status and outlook", *Reports on Progress in Physics*, vol. 68, pp. 2267–2336, 2005.

[RYL 11] RYLE J.P., LI D., SHERIDAN J.T., "Feasibility investigation of using tunable infrared communications laser for multiwavelength digital holographic Laplacian reconstruction", *Optical Engineering*, vol. 50, no. 10, p. 105801, 2011.

[SCH 07] SCHNEIDER H., LIU H.C., *Quantum Well Infrared Photodetectors: Physics and Applications*, Springer Series in Optical Sciences, Springer, Berlin Heidelberg, vol. 126, 2007.

[SIM 70] SIMPSON W.A., DEEDS W.E., "Real-time visual reconstruction of infrared holograms", *Applied Optics*, vol. 9, no. 2, pp. 499–501, 1970.

[SKO 03] SKOTHEIM Ø., "HoloVision: a software package for reconstruction and analysis of digitally sampled holograms", *Proceedings of SPIE*, vol. 4933, pp. 311–316, 2003.

[STO 90] STOVER J.C., *Optical Scattering – Measurement and Analysis*, McGraw-Hill Inc, 1990.

[STO 12] STOYKOVA E., YARAS F., KANG H., *et al.*, "Visible reconstruction by a circular holographic display from digital holograms recorded under infrared illumination", *Optics Letters*, vol. 37, no. 15, pp. 3120–3122, 2012.

[SUT 00] SUTKOWSKI M., KUJAWINSKA M., "Application of liquid crystal (LC) devices for optoelectronic reconstruction of digitally stored holograms", *Optics and Lasers in Engineering*, vol. 33, no. 3, pp. 191–201, 2000.

[SVE 84] SVELTO O., *Principle of Lasers*, Plenum, New York, 1984.

[TIS 04] TISSOT J.L., "IR detection with uncooled focal plane arrays. State-of-the-art and trends", *Opto-Electronics Review*, vol. 12, no. 1, pp. 105–109, 2004.

[VAN 07] VANDENRIJT J.-F., GEORGES M., "Infrared electronic speckle pattern interferometry at 10 µm", *Proceedings of SPIE,* vol. 6616, p. 66162Q, 2007.

[VAN 09] VANDENRIJT J.-F., THIZY C., ALEXEENKO I., *et al.*, "Electronic speckle pattern interferometry at long infrared wavelengths, Scattering requirements", *Fringe 2009 – 6th International Workshop on Advanced Optical Metrology*, Springer, pp. 596–599, 2009.

[VAN 10] VANDENRIJT J.-F., GEORGES M., "Electronic speckle pattern interferometry with microbolometer arrays at 10.6 µm", *Applied Optics,* vol. 49, no. 27, pp. 5067–5075, 2010.

[VAN 13] VANDENRIJT J.-F., THIZY C., ALEXEENKO I., *et al.*, "Mobile speckle interferometer in the long-wave infrared for aeronautical nondestructive testing in field conditions", *Optical Engineering*, vol. 52, no. 10, p. 101903, 2013.

[VAN 14a] VANDENRIJT J.-F., THIZY C., QUEECKERS P., *et al.*, "Long-wave infrared digital holographic interferometry with diffuser or point source illuminations for measuring deformations of aspheric mirrors", *Optical Engineering*, vol. 53, no. 11, p. 112309, 2014.

[VAN 14b] VANDENRIJT J.-F., THIZY C., STOCKMAN Y., *et al.*, "Digital holographic interferometry in the long-wave infrared for the testing of large aspheric space reflectors", *Fringe 2013 – 7th International Workshop on Advanced Optical Metrology*, Springer, pp. 925–928, 2014.

[VAN 14c] VANDENRIJT J.-F., THIZY C., GEORGES M.P., "Vibration analysis by speckle interferometry with CO_2 lasers and microbolometers arrays", *Imaging and Applied Optics 2014*, OSA Technical Digest (online), paper DTh4B.8, 2014, available at http://www.opticsinfobase.org/abstract.cfm?URI=DH-2014-DTh4B.8.

[YAM 03] YAMAGUCHI I., "Fundamentals and applications of speckle", *Proceedings of SPIE*, vol. 4933, pp. 1–8, 2003.

[YAO 12] YAO Y., HOFFMAN A.J., GMACHL C.F., "Mid-infrared quantum cascade lasers", *Nature Photonics*, vol. 6, pp. 432–439, 2012.

7

Full Field Holographic Vibrometry at Ultimate Limits

7.1. Introduction

Digital holography is a fast-growing research field that has drawn increasing attention. The main advantage of digital holography is that, contrary to holography with photographic plates, the holograms are recorded by a charge-coupled device (CCD), and the image of the object is digitally reconstructed by a computer, avoiding photographic processing [SCH 94]. To extract two quadratures of the holographic signal (i.e. to get the amplitude and the phase of the optical signal), two main optical configurations have been developed: off-axis and phase-shifting.

Off-axis holography is the oldest and the simplest configuration adapted to digital holography. In that configuration, the reference or local oscillator (LO) beam is angularly tilted with respect to the object observation axis. It is then possible to record, with a single hologram, the two quadratures of the object complex field [SCH 94]. However, the object field of view is reduced, since we must avoid the overlapping of the image with the conjugate image alias. Phase-shifting digital holography [YAM 97] makes it possible to obtain phase information on

Chapter written by Nicolas VERRIER, Michael ATLAN and Michel GROSS.

the whole camera area by recording several images with a different phase for the reference (called LO) beam. It is then possible to obtain the two quadratures of the field in an on-axis configuration even though the conjugate image alias and the true image overlap, because aliases can be removed by taking image differences. In a typical phase-shifting holographic setup, the phase of the reference is shifted by moving a mirror with a Piezoelectric Transductor (PZT).

However, there is a big demand for full-field vibration measurements, particularly in industry. Different holographic techniques are able to image and analyze such vibrations. Double-pulse holography [PED 95, PED 97] records a double-exposure hologram with time separation in the 1...1,000 μs range, and measures the instantaneous velocity of a vibrating object from the phase difference. The method requires a quite costly double-pulse ruby laser system, whose repetition rate is low. Multi-pulse holography [PED 98] is able to analyze transient motions, but the setup is still heavier (four pulses laser, three cameras).

The development of fast Complementary Metal Oxide Semiconductor (CMOS) camera makes it possible to analyze vibration efficiently by triggering the camera on the motion in order to record a sequence of holograms that allows us to track the vibration of the object as a function of time [PED 06, FU 07]. The analysis of the motion can be done by phase difference or by Fourier analysis in the time domain. The method requires a CMOS camera, which can be costly. It is also limited to low-frequency vibrations, since a complete analysis of the motion requires a camera frame rate higher than the vibration frequency, because the bandwidth BW of the holographic signal, which is sampled at the camera at angular frequency ω_{CCD}, must be lower than corresponding Nyquist–Shannon limit: $BW < \frac{1}{2}\,\omega_{CCD}$.

For a periodic vibration motion, the bandwidth BW is close to zero. Thus, measurements can be done using much slower cameras. Powell and Stetson [PIC 05b] have shown, for example, that a harmonically vibrating object yields alternate dark and bright fringes, whose analysis yields informations on the vibration motion. Picard *et al.* [PED 95] have simplified the processing of the data by performing time-averaged holography with a digital CCD camera. Time-averaged holography has no limit in vibration frequency and it involves neither a costly laser system, nor an expensive CMOS fast camera. Although the time-averaging method gives a way for determining the amplitude of vibration [PIC 03], quantitative measurement remains quite difficult for low and high vibration amplitudes.

Heterodyne holography [LE 00, LE 01] is a variant of phase-shifting holography in which the frequency, phase and amplitude of both reference and signal beams are controlled by acousto-optic modulators (AOM). Heterodyne holography is thus extremely versatile. By shifting the frequency ω_{LO} of the LO beam with respect to the frequency ω_0 of illumination, it is, for example, possible to detect the holographic signal at a frequency ω different than illumination ω_0. This ability will be extremely useful to analyze vibration, since heterodyne holography can detect selectively the signal that is scattered by the object on a vibration sideband of frequency $\omega_m = \omega_0 + m\omega_A$, where ω_A is the vibration frequency and m is the integer index.

In this chapter, we will first present in section 7.2 heterodyne holography and its advantages in sections 7.2.1 and 7.2.2. Then, in section 7.3, we will apply heterodyne holography to vibration analysis. We will show in section 7.3.2 how heterodyne holography can be used to detect the vibration sidebands, in section 7.3.4 how this sideband holography can be combined with stroboscopic techniques to record instantaneous velocity maps of motion, and in sections 7.3.3 and 7.3.5 how it can retrieve both small and large vibration amplitudes.

7.2. Heterodyne holography

Let us first describe heterodyne holography. An example of setup is shown in Figure 7.1. The object is a U.S. Air Force (USAF) resolution target whose hologram is recorded in transmission geometry (the target is back-illuminated). The camera records the interference of the signal field $\mathcal{E}(t)$ of optical angular frequency ω_0, with the reference (or LO field) $\mathcal{E}_{LO}(t)$ of optical frequency:

$$\omega_{LO} = \omega_0 - \Delta\omega \qquad [7.1]$$

where $\Delta\omega$ is the frequency shift. Let us introduce the slowly varying complex fields E and E_{LO}.

$$\mathcal{E}(t) = Ee^{j\omega_0 t} + E^* e^{-j\omega_0 t} \quad [7.2]$$

$$\mathcal{E}_{LO}(t) = E_{LO}e^{j\omega_{LO} t} + E_{LO}^* e^{-j\omega_{LO} t}$$

where E^* and E_{LO}^* are the complex conjugates of E and E_{LO}, and j is the imaginary unit. The camera signal I is proportional to the intensity of the total field $|\mathcal{E}(t) + \mathcal{E}_{LO}(t)|^2$. We thus have:

$$I = \left|E_{LO}\, e^{j\omega_{LO} t} + E\, e^{j\omega_0 t}\right|^2 \quad [7.3]$$

$$= |E_{LO}|^2 + |E|^2 + E_{LO}^* E\, e^{+j\Delta\omega t} + E_{LO} E^*\, e^{-j\Delta\omega t}$$

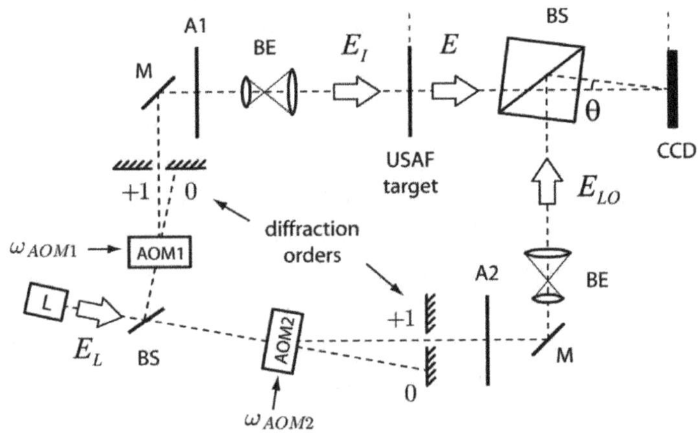

Figure 7.1. *Typical heterodyne holography setup. L: main laser; BS: beam splitter; AOM1, AOM2: acousto-optic modulators; BE: beam expander; M, mirror; A1, A2: light attenuators; USAF, transmission USAF 1951 test pattern; CCD: camera;* E_L, E_I, E_{LO}, *and* E: *laser, illumination, reference (i.e. local oscillator – LO) and object fields;* $\omega_{AOM1/2}$: *driving frequencies of the acousto-optic modulators AOM1 and AOM2;* θ: *off-axis angular tilt*

Since the camera signal is slowly varying, we have neglected in equation [7.3] the fast varying terms (which evolve at frequency $\simeq 2\omega_0$). Moreover, to simplify the present discussion, we have not considered the spatial variations of I, E and E_{LO} with x and y, in particular the spatial variations that are related to the off-axis tilt angle θ of Figure 7.1. In equation [7.3], $E_{LO}^{*}E\ e^{-j\Delta\omega t}$ is the $+1$ grating order term that contains the useful information (since this term is proportional to E). The others terms $|E_{LO}|^2 + |E|^2$ (zero grating order term) and $E_{LO}E^{*}\ e^{+j\Delta\omega t}$ (-1 grating order term or twin image term) are unwanted terms that must be canceled. To filter off these terms, and to select the desired $+1$ grating order signal, four-phase detection is done. The AOMs driving angular frequencies ω_{AOM1} and ω_{AOM2} are tuned to have:

$$\Delta\omega = \omega_{AOM2} - \omega_{AOM1} = \omega_{CCD}/4 \qquad [7.4]$$

where ω_{CCD} is the angular frequency of the camera frame rate. Let us consider four successive camera frames I_0, I_1 ...I_3 that are recorded at times $t = 0, T$ ... $3T$ with $T = 2\pi/(\omega_{CCD})$. For these frames, the phase factor $e^{+j\Delta\omega t}$ is equal to 1, j, -1 and $-j$. Thus, we get:

$$\begin{aligned} I_0 &= |E_{LO}|^2 + |E|^2 + \ E_{LO}^{*}E + \ E_{LO}E^{*} \qquad [7.5] \\ I_1 &= |E_{LO}|^2 + |E|^2 + j\ E_{LO}^{*}E - j\ E_{LO}E^{*} \\ I_2 &= |E_{LO}|^2 + |E|^2 - \ E_{LO}^{*}E - \ E_{LO}E^{*} \\ I_3 &= |E_{LO}|^2 + |E|^2 - j\ E_{LO}^{*}E + j\ E_{LO}E^{*} \end{aligned}$$

By linear combination of four frames, we get the four-phase hologram H that respects the demodulation equation:

$$\begin{aligned} H &= (I_0 - I_2) + j(I_1 - I_3) \qquad [7.6] \\ &= 4\ E_{LO}^{*}E \end{aligned}$$

As desired, the four-phase demodulation equation yields a quantity H that is proportional to the signal complex field E. Note that the coefficient $4E_{LO}$ is supposed to be known and does not depend on the object.

Heterodyne holography exhibits several advantages compared to other holographic techniques:

1) The phase shift is very accurate.

2) The holographic detection is shot-noise limited.

3) Since the holographic detection is done somewhere near ω_{LO} (depending on the demodulation equation that is chosen), it is possible to perform the holographic detection with any frequency shift with respect to the object illumination angular frequency ω_0.

The first advantage will be discussed in section 7.2.1, the second advantage will be discussed in section 7.2.2, and the third advantage, which is the heart of the sideband holographic technique used to analyze vibration, will be discussed in section 7.3.2.

7.2.1. *Accurate phase shift and holographic detection bandwidth*

In the typical heterodyne holography setup in Figure 7.1, the signals that drive the AOMs are generated by frequency synthesizers, phase-locked with a common 10 MHz clock. The phase shift $\Delta\varphi = \Delta\omega T$ (that is equal to $\pi/2$ in the four-phase demodulation case) can thus be adjusted with quartz accuracy.

To illustrate this accurate phase shift, holograms of a USAF target have been recorded with the Figure 7.1 setup, while sweeping the LO frequency ω_{LO}. Figure 7.2 shows USAF reconstructed images which are obtained for different values of the frequency shift $\omega_0 - \omega_{LO}$ [ATL 07]:

1) For $\omega_0 - \omega_{LO} = \omega_{CCD}/4 = +2.5$ Hz, the image reconstructed is sharp and corresponds to the $+1$ grating order. However, the -1 grating order signal is very low; the magnitude of the –1 parasitic twin image is negligible in front of the $+1$ contribution. Image and zoom of the image are shown in Figures 7.2(a) and (c).

2) For $\omega_0 - \omega_{LO} = -\omega_{CCD}/4 = -2.5$ Hz, the image reconstructed is blurred and corresponds to the –1 grating order twin image. The magnitude of the $+1$ grating order image is negligible in front of its -1 counterpart. Image and zoom of the image are shown in Figures 7.2 (b) and (d).

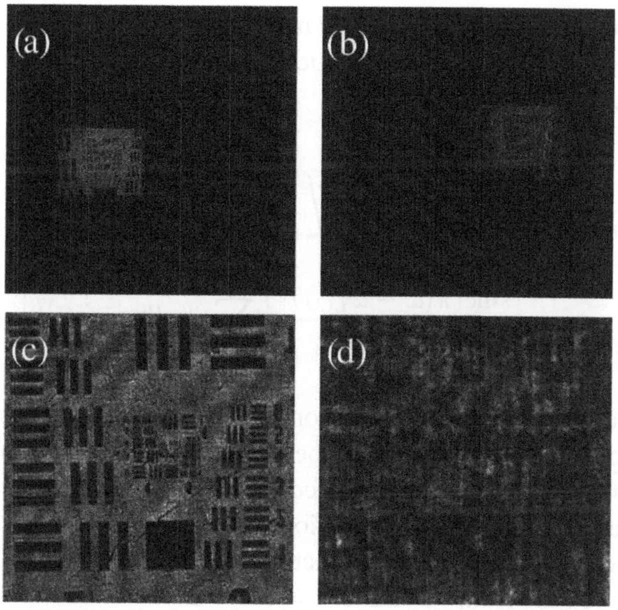

Figure 7.2. *USAF target reconstructed images (a,b) displayed with linear gray scale for the reconstructed field intensity* $|E|^2$ *with* $\omega_0 - \omega_{LO} = \omega_{CCD}/4 = +2.5$ *Hz a) and* $\omega_0 - \omega_{LO} = -\omega_{CCD}/4 = -2.5$ *Hz b). Zooms (c,d) of the bright zones of the a) and b) images. Images (a,b) and zoom (c,d) correspond to the* $m = +1$ *(a,c) and* $m = -1$ *(b,d) grating orders*

We have measured the total field energy $U_\pm$ in both $+1$ and -1 grating order reconstructed images:

$$U_\pm - \sum_{x,y} |E(x,y)|^2 \qquad [7.7]$$

where $\sum_{x,y}$ is the sum over the pixels of the reconstructed field image $E(x,y)$ corresponding to the ± 1 grating order zones. As shown in Figure 7.3, the energy of the signal measured in the $+1$ grating order (i.e. U_+) is maximum for $\omega_0 - \omega_{LO} = \omega_{CCD}/4 = +2.5$ Hz, and null for $\omega_0 - \omega_{LO} = -2.5$ Hz. Similarly, U_- is maximum for $\omega_0 - \omega_{LO} = -2.5$ Hz, and null for $\omega_0 - \omega_{LO} = +2.5$ Hz. By adjusting $\omega_0 - \omega_{LO}$, it is then possible to select the grating order that is detected.

For four-phase detection with an LO of frequency ω_{LO}, the total energy $U_{\pm} = |E|^2$ detected at frequency ω in the ± 1 grating order can easily be calculated [VER 10]:

$$U_{\pm}(\omega - \omega_{LO}) = \left| \frac{1}{4T'} \sum_{n=0}^{n=3} (\pm j)^n \int_{t=nT}^{nT+T'} dt\ e^{\pm j(\omega - \omega_{LO})t} \right|^2 \qquad [7.8]$$

$$= \left| \frac{\mathrm{sinc}(\pi(\omega - \omega_{LO})T')}{4} \sum_{n=0}^{n=3} j^n e^{jn(\omega - \omega_{LO})T} \right|^2$$

where $\mathrm{sinc}(x) = \sin x / x$. In equation 7.8, T is the frame period and T' is the exposure time. Thus, the coefficient $1/4T'$ is a normalization factor. However, the factors $(\pm j)^n$ correspond to the coefficients of the demodulation equation (see equation [7.6]), and $e^{\pm j(\omega - \omega_{LO})t}$ is the interference instantaneous phase factor, which must be integrated over the exposure time from $t = nT$ to $t = nT + T'$.

In the experiment of Figure 7.3, we have $T = T' = 2\pi/\omega_{CCD} =$ 100 ms. We have shown the field energy $U_{\pm}$ given by equation 7.8 as a function of $\omega - \omega_{LO}$ in Figure 7.3 (solid gray lines). As shown in the figure, experiment agrees with the theoretical curve of equation 7.8. The shape of the curves represents here the frequency response spectrum of the holographic device considered as a detector. For the $+1$ grating order, detection is centered at frequency $\omega = \omega_{LO} + \omega_{CCD}/4$. The measurement bandwidth BW is 2.5 Hz. It is equal to the inverse of the measurement time of four frames, i.e. $\mathrm{BW} = 1/4T$. It illustrates the coherent character (in time) of holographic detection.

It is possible to increase the selectivity of the holographic coherent detection by increasing the measurement time, i.e. by increasing the number of frames n_{max} used for demodulation. In that case, equations [7.6] and [7.8] must be replaced by similar equations involving n_{max} frames in place of four. We have:

$$H = \sum_{n=0}^{n_{max}-1} j^n\ I_n \qquad [7.9]$$

and

$$U_{\pm}(\omega - \omega_{LO}) = \left| \frac{1}{n_{max}T} \sum_{n=0}^{n_{max}-1} (\pm j)^n \int_{t=nT}^{nT+T'} dt\ e^{\pm j(\omega-\omega_{LO})t} \right|^2$$
$$= \left| \frac{\text{sinc}(\pi(\omega - \omega_{LO})T')}{n_{max}} \sum_{n=0}^{n_{max}-1} j^n e^{jn(\omega-\omega_{LO})T} \right|^2 \quad [7.10]$$

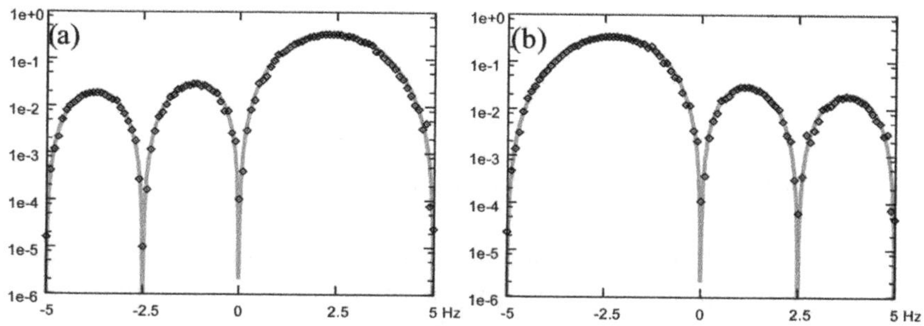

Figure 7.3. *Field energy of the a) +1 and b) −1 grating orders. The vertical axis is in logarithmic scale arbitrary units. The points correspond to the experimental data obtained by sweeping the LO frequency with* $(\omega_0 - \omega_{LO}) = -5$ *to* $+5$ *Hz with 0.1 Hz increments. The solid gray curves are a)* $+U_+(\omega - \omega_{LO})$ *and b)* $+U_-(\omega - \omega_{LO})$ *given by equation 7.8 in the same frequency range:* $(\omega - \omega_{LO}) = -5$ *to* $+5$ *Hz*

Figure 7.4 shows the detection frequency spectrum U_+ for $n_{max} = 4$, 8 and 16. As shown in the figure, the detection bandwidth BW decreases with n_{max}. It is equal to:

$$\text{BW} = \frac{1}{n_{max}T} \quad [7.11]$$

The noise, uniformly distributed in frequency, is expected to decrease accordingly. We will show in section 7.2.2 that very high detection sensitivity can be obtained by holographic detection with large number of frames n_{max}.

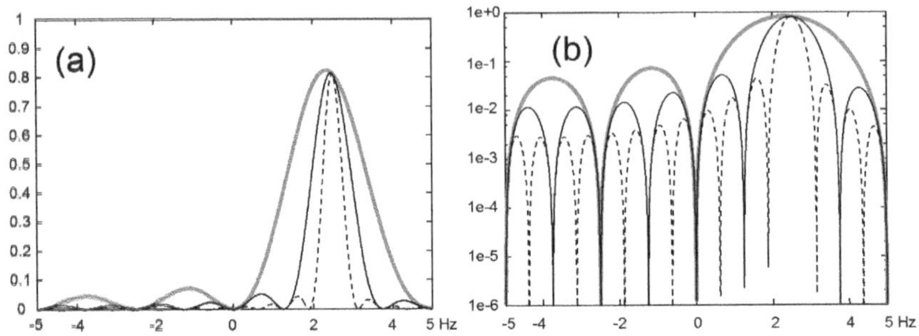

Figure 7.4. *Theoretical field energy $U_+(\omega - \omega_{LO})$ given by equation [7.8] plotted in linear a) and logarithmic scale b) as a function of $(\omega - \omega_{LO})$ in Hz units for the +1 grating order. The number of frame is $n_{max} = 4$ (solid gray line), $n_{max} = 8$ (solid black line) and $n_{max} = 16$ (dashed black line)*

7.2.2. *Shot noise holographic detection*

Because the holographic signal results in the interference of the object signal complex field E with a reference (LO) complex field E_{LO} whose amplitude can be much larger (i.e. $E_{LO} \gg E$), the holographic detection benefits from "heterodyne" or "holographic" gain (i.e. $|EE^*_{LO}| \gg |E|^2$), and is thus well suited for the detection of weak signal fields E. Holographic detection can reach the theoretical limit of noise which corresponds to a noise equivalent signal of one photoelectron per pixel during the total measurement time [GRO 07, GRO 08, VER 10, LES 13].

To illustrate this point, let us consider the interference of a weak signal of about one photoelectron per pixel and per camera frame, with a large LO reference signal of $N = 10^4$ photoelectrons (see Figure 7.5). In this example, the signal intensity is $|E|^2 = 1$, while the holographic interference term $|EE^*_{LO}| = 100$ is much larger. The detected signal is $|E_{LO} + E|^2 \simeq 10^4$ photoelectrons.

Because of the quantum nature of the process involved in digital holography (laser emission, photodetection, etc.), the detected signal in photoelectron units is random Gaussian integer number, whose average value is $N = 10^4$, and whose standard deviation is $\sqrt{N} = 100$. These fluctuations of the number of photoelectrons are called shot noise.

Here, with one photoelectron of signal, the heterodyne signal is 100 photoelectrons and the noise is 100 too. Thus, the noise equivalent signal (for the energy $|E|^2$) is one photoelectron per pixel and per frame.

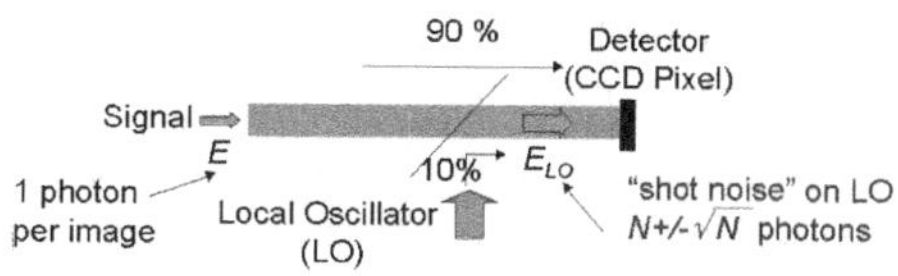

Figure 7.5. *Heterodyne detection of one photon per pixel of signal with N photon local oscillator*

Let us now study how shot noise varies with the number of frames n_{max} used for detection. As in any detection process, the noise in energy is proportional to the measurement time and to the detection bandwidth BW. Since the measurement time is $n_{max}T$, and since the detection bandwidth is BW $= 1/n_{max}T$ (see equation [7.11]), the shot noise in energy does not depend on n_{max}. The shot-noise equivalent signal is thus the same as for one frame. It remains equal to one photoelectron per pixel whatever the number of frames n_{max} and measurement time $n_{max}T$ are.

Let us now discuss the ability to reach this shot-noise optimal sensitivity in real-life holographic experiments. Since we consider implicitly a very weak signal, the noises that must be considered are:

1) the read noise and dark current of the camera;

2) the quantization noise of the camera A/D converter;

3) the technical noise of the LO beam;

4) the LO beam shot noise, which yields the theoretical noise limit.

For a typical camera, the full well capacity is $2 * 10^4$ photoelectrons, and a good practice is to work with $N = 10^4$ photoelectrons for the LO. Shot noise on the camera signal (100 photoelectrons) is thus much larger than the camera read noise (1–20 photoelectrons) and the camera dark current (a few photoelectrons per second). If the camera is 12 bit, the full well capacity corresponds to $\sim 2^{12}$ digital count (DC). As a result, the quantization noise ($\sim$ 7 photoelectrons) can be neglected too.

Let us discuss the technical noise of the LO beam, and the way to filter it off:

1) We have displayed in Figure 7.6 examples of Fourier space hologram $\tilde{H}(k_x, k_y) = \mathrm{FFT}(H(x, y))$ (where FFT is the Fourier transform operator). The signal fields yield the interference term EE^*_{LO} which is located in the left-hand side of the Fourier space images of Figures 7.6(a) and (b). It corresponds to the grating order $+1$. The LO signal $|E_{LO}|^2$ is located in the center of the Fourier space. It is visible in Figure 7.6(a) which is obtained with one phase hologram. As shown in the figure, the LO signals are separated in the Fourier space because of the off-axis configuration. It is then possible to filter off the parasitic LO signal by a proper spatial filtering in the Fourier space as shown by Cuche *et al.* [CUC 00]. As noticed, this operation filters off both the LO signal and the LO technical noise.

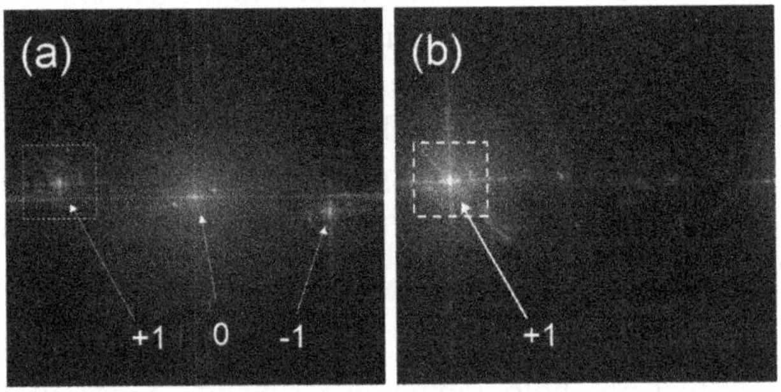

Figure 7.6. *Fourier space hologram intensity (i.e. $|\tilde{H}(k_x, k_y)^2$) displayed in arbitrary log scale for one phase ($H = I_0$) and four-phases ($H = (I_0 - I_2) + j(I_1 - I_3)$ detection*

2) On the other hand, because of phase-shifting, the signal and LO fields E and E_{LO} evolve at two different time frequencies ω and ω_{LO} with $\omega = \omega_{LO} + \omega_{CCD}/4$. It is then possible to filter off the LO signal in time. This is done by the four-phases demodulation process since the LO term $|E_{LO}|^2$ vanishes in equations [7.6] and [7.10]. This filtering in time is illustrated in Figure 7.6, since the zero grating order signal that is large for single-phase off-axis holography (Figure 7.6(a)) roughly vanishes in the four-phase case (Figure 7.6(b)). Time filtering is also illustrated in Figures 7.3 and 7.4 and equations [7.6] and [7.10], since

the detected energy $U_{+/-}$ vanishes for $\omega - \omega_{LO} = 0$, i.e. for detection at LO frequency ω_{LO}. Here again, we filter off both the LO signal and the LO technical noise.

By combining off-axis and phase-shifting, it is possible to apply to a double filtering (in space and time) which filters off the LO technical noise very effectively. Then, we can get the shot-noise ultimate sensitivity that corresponds to a noise equivalent signal of one photon per pixel whatever the measurement time is.

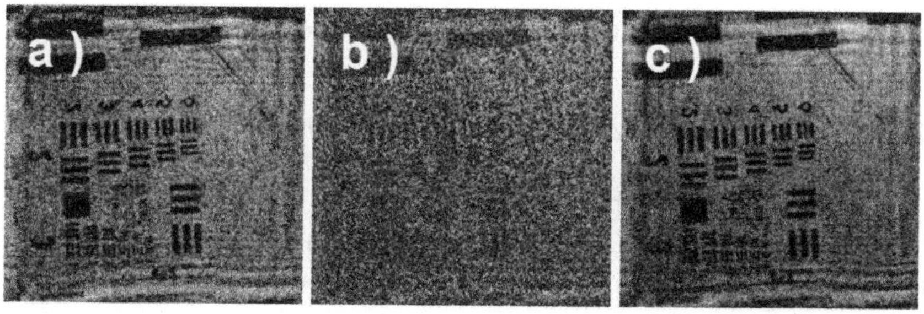

Figure 7.7. *Reconstructed image of a USAF target in dim light. Holograms are recorded by four-phase heterodyne holography with $T = 100$ ms, and $n_{max} = 600$ (a,b) and $n_{max} = 6{,}000$ (c). The coherent measurement time $n_{max}T$ is thus 1 min (a,b) and 10 min (c). Illumination is adjusted so that the USAF signal integrated is 1 photon per pixel over the whole measurement time $n_{max}T$ in a). Holograms (b) and (c) are recorded by adding a neutral density filter D = 1.0 on illumination. The USAF signal is thus 0.1 photon per pixel in b) and 1 photon per pixel in c). The display is made in arbitrary linear scale for intensity*

Figure 7.7 illustrates heterodyne holography ability of imaging an object (here a USAF target) in dim light with shot-noise-limited sensitivity. In Figure 7.7(a), the measurement time is 1 min and the illumination power is adjusted in such a way that the USAF signal is about one photon per pixel in 1 min. The visual quality of USAF image is quite good. We can say that the signal-to-noise ratio (SNR) is about 1. In Figure 7.7(b), the measurement time is the same, but illumination is divided by 10 by using a neutral filter of density $D = 1.0$. SNR is low, and we cannot see the grooves of the USAF target. Figure 7.7(c) is obtained with the same illumination level as Figure 7.7(b) (i.e. with neutral density filter) but the measurement time is multiplied by 10 (i.e. 10 min in place of 1 min). The visual quality of the reconstructed image is the same as for Figure 7.7(a) with SNR about 1. This experiment shows that the reconstructed image quality depends on the total amount of signal, and does not depend on the time needed to get that amount of signal: to get SNR $\sim$ 1, we need approximately one photon per pixel for any measurement time.

7.3. Holographic vibrometry

Let us now apply heterodyne holography to vibration analysis.

7.3.1. *Optical signal scattered by a vibrating object*

Consider a point of the object (e.g. a clarinet reed) that is studied by heterodyne holography (see Figure 7.8), vibrating at frequency ω_A with amplitude z_{max}. The displacement z along the out of plane direction is:

$$z(t) = z_{max} \sin \omega_A t \qquad [7.12]$$

In backscattering geometry, this corresponds to a phase modulation $\varphi(t)$ of the signal:

$$\begin{aligned} \varphi(t) &= 4\pi z(t)/\lambda \\ &= \Phi \sin \omega_A t \end{aligned} \qquad [7.13]$$

where λ is the optical wavelength and Φ is the amplitude of the phase modulation of the signal at angular frequency ω_A:

$$\Phi = 4\pi z_{max}/\lambda \qquad [7.14]$$

Let us define the slowly varying complex amplitude $E(t)$ of the field $\mathcal{E}(t)$ scattered by the vibrating object. We have:

$$\mathcal{E}(t) = E(t)e^{j\omega_0 t} + \text{c.c.} \qquad [7.15]$$

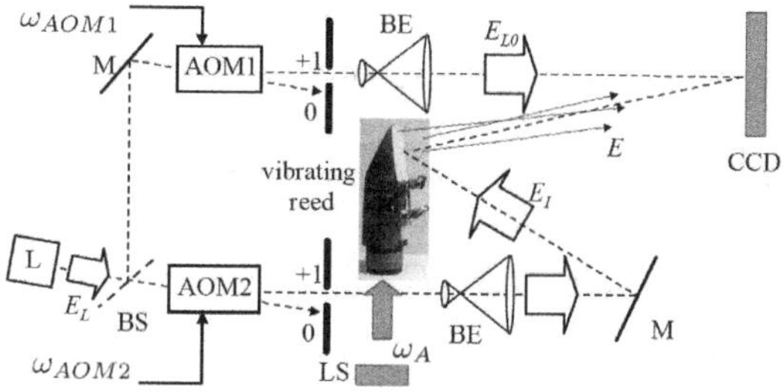

Figure 7.8. *Heterodyne holography setup applied to analyze vibration of a clarinet reed. L: main laser; AOM1, AOM2: acousto-optic modulators; M: mirror; BS: beam splitter; BE: beam expander; CCD: camera; LS: loudspeaker exiting the vibrating clarinet reed at frequency* $\omega_A/2\pi$

Because of the Jacobi–Anger expansion, we get:

$$\begin{aligned} E(t) &= E\, e^{j\varphi(t)} = E\, e^{j\Phi \sin \omega_A t} \\ &= E \sum_m J_m(\Phi)\, e^{jn\omega_A t} \end{aligned}$$

where E is the complex amplitude without vibration, and J_m is the mth-order Bessel function of the first kind, with $J_{-m}(z) = -1^m J_m(z)$ for integer m and real z. The scattered field $\mathcal{E}(t)$ is then the sum of the carrier and sideband field components $\mathcal{E}_m(t)$ of frequency ω_m, where m is the sideband index with:

$$\begin{aligned} \mathcal{E}(t) &= \sum_{m=-\infty}^{+\infty} \mathcal{E}_m(t) \\ \mathcal{E}_m(t) &= E_m e^{j\omega_m t} + E_m^* e^{-j\omega_m t} \\ \omega_m &= \omega_0 + m\omega_A \end{aligned} \qquad [7.16]$$

where E_m is the complex amplitude of the field component $\mathcal{E}_m(t)$. Equation [7.16] yields:

$$E_m = J_m(\Phi)\, E \qquad [7.17]$$

Figure 7.9 presents the distribution of the field energy on the sideband components $|E_m|^2$. If the amplitude of modulation Φ is low, most of the energy is on the carrier $|E_0|/|E|^2 \simeq 1$, and energy $|E_m|^2$ decreases rapidly with the sideband index m. If the amplitude Φ is large, the energy of the carrier is low $|E_0|/|E|^2 \ll 1$, while energy is distributed over many sidebands $|E_m|^2$.

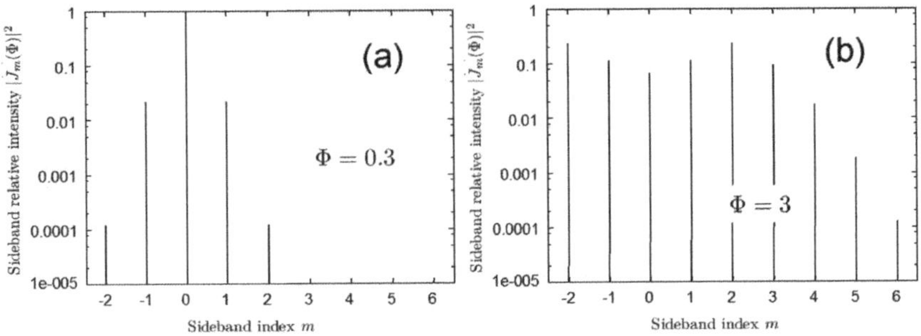

Figure 7.9. *Relative amplitude $|E_m|^2/|E|2 = |J_m(\Phi)|^2$ of the sideband component m for an amplitude modulation of the phase equal to $\Phi = 0.3$ a) and $\Phi = 3.0$ b) rad*

7.3.2. *Selective detection of the sideband components E_m: sideband holography*

Heterodyne holography which is able to perform the holographic detection with any frequency shift $\omega - \omega_0$ with respect to the object illumination angular frequency ω_0 is well suited to detect the vibration sideband components E_m. To selectively detect by four-phase demodulation the sideband m of frequency ω_m, the frequency ω_{LO} must be adjusted to fulfill the condition:

$$\begin{aligned} \omega_{LO} &= \omega_m - \omega_{CCD}/4 \\ &= \omega_0 + m\omega_A - \omega_{CCD}/4 \end{aligned} \quad [7.18]$$

Figure 7.10 shows images obtained by detecting different sideband m of a clarinet reed [JOU 09a]. The clarinet reed is attached to a clarinet mouthpiece and its vibration is driven by a sound wave propagating inside the mouthpiece, as in playing conditions, but the sound wave is created by a loudspeaker excited at frequency ω_A and has a lower intensity than inside a clarinet. The excitation frequency is adjusted to be resonant with the first flexion mode (2,143 Hz) of the reed.

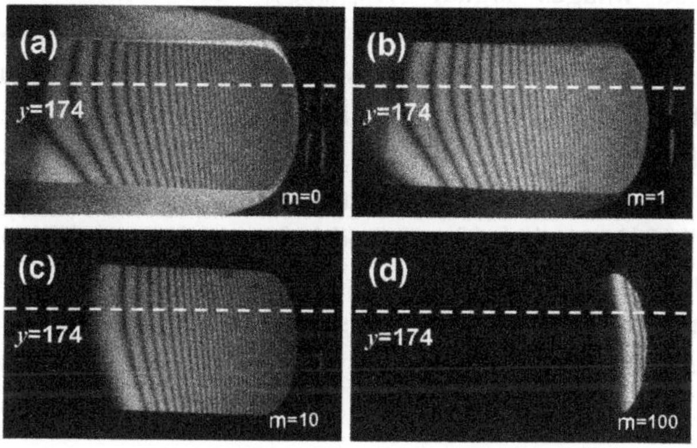

Figure 7.10. *Reconstructed holographic images of a clarinet reed vibrating at frequency $\omega_A/2\pi = 2143$ Hz perpendicularly to the plane of the figure. a) The carrier image obtained for $m = 0$. b)–d) The frequency sideband images, respectively, for $m = 1$, $m = 10$ and $m = 100$. A logarithmic gray scale has been used*

Figure 7.10(a) is obtained at the unshifted carrier frequency ω_0. It corresponds to an image obtained by time-averaging holography [PED 95]. The left side of the reed is attached to the mouthpiece, and the amplitude of vibration is larger at the tip of the reed on the right side; in this region, the fringes become more closer and difficult to count. The mouthpiece is also visible, but without fringes since it does not vibrate. Similar images of clarinet reeds have been obtained in [DEM 04, PIC 05a], with more conventional techniques and lower image quality. Figures 7.10(b)–(d) show images obtained for the sidebands $m = 1$, 10 and 100. As expected, the non-vibrating mouthpiece is no longer visible. Figure 7.10(b) shows the $m = 1$ sideband image, with J_1 fringes that are slightly shifted with respect to those of J_0. Figure 7.10(c) shows the image of sideband $m = 10$ and Figure 7.10(d) shows the image of sideband $m = 100$. The left side region of the image remains dark because, in that region, the vibration amplitude is not sufficient to generate these sidebands, $J_m(z)$ being evanescent for $z < m$.

To quantitatively visualize the vibration amplitude Φ, cuts of the reconstructed images signal $|E_m(x,y)|^2$ are made for different sideband orders m along the horizontal line $y = 174$. This value has been chosen because it corresponds to a region where the fringes are orthogonal to the y-axis. To build the central part of Figure 7.11, the position of the antinodes of $|E_m(x,174)|^2$ are put in correspondence with the antinodes of $|J_m(\Phi)|^2$. Correspondence is made for $m = 0, 1, 5, 10, \ldots 100$. Note that this method is insensitive to inhomogeneities in the illumination zone. Therefore, no normalization is required. The curve seen in the central zone of Figure 7.11 represents the amplitude of phase oscillation Φ in radian as a function of pixel index along x for $y = 174$.

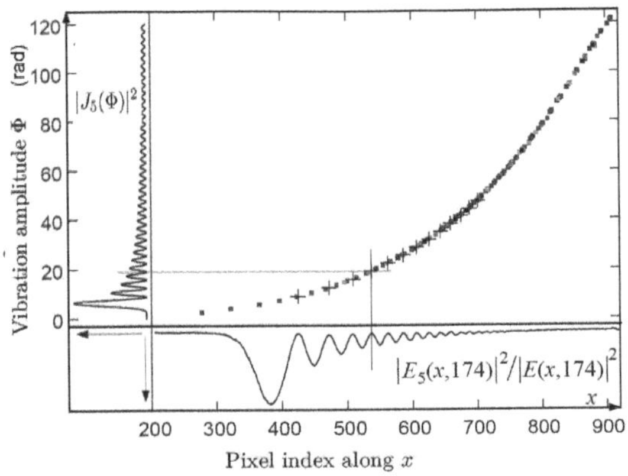

Figure 7.11. *A slice of the data along the $y = 174$ line is used in this figure; the x horizontal axis gives the pixel index (100 pixels correspond to 3.68 mm), and the vertical axis gives the vibration amplitude Φ. The lower part of the figure shows the normalized signal $|E_m(x)|^2/|E(x)|^2$ (where $E(x)$ is obtained loudspeaker off) for $m = 5$, with a downward axis. The left part shows the corresponding square of the Bessel function $|J_5(\Phi)|^2$ with a leftward axis. The zeros of the two curves are put in correspondence, which provides the points in the central figure. Similar correspondences are made harmonic order $m = 0, 1, 5, 10...100$. Different gray densities are used for different m. The crosses correspond to $m = 5$. The juxtaposition of the points for all values of m gives an accurate representation of the amplitude of vibration Φ as a function of x*

REMARK 7.1.– In a typical heterodyne holography setup, digital synthesizers drive the AOM at ω_{AOM1} and ω_{AOM2}, the camera at ω_{CCD} and the vibration frequency at ω_A. These synthesizers use a common 10 MHz reference frequency and are driven by the computer. It is then possible to automatically sweep ω_A, and ω_{AOM1} (or ω_{AOM2}), in order to fulfill equation [7.18] so that detection remains ever tuned on a given sideband. Figure 7.12 shows an example [TAI 14]. A series of 26×7 images of a clarinet reed are obtained on sideband $m = 1$ by sweeping the frequency ω_A from 1.4 up to 20 kHz by steps of 25 cents. The amplitude of the excitation signal is exponentially increased in the range 1.4–4 kHz, from 0.5 to 16 V, and then kept constant at 16 V up to 20 kHz. This crescendo limits the amplitude of vibration of the first two resonances of the reed. The different vibration modes of the reed can easily be recognized on the reconstructed reed images of Figure 7.12.

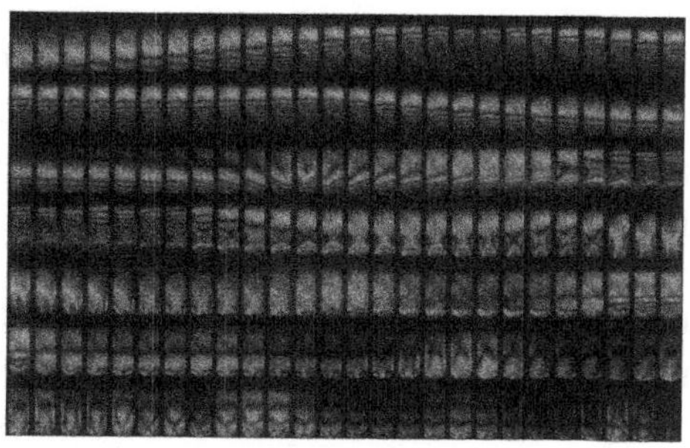

Figure 7.12. *Clarinet reed reconstructed images obtained on sideband $m = 1$. Frequency ω_A is swept from 1.4 up to 20 KHz, and images are displayed from left to right and top to bottom (26×7 images). The display is made with arbitrary gray scale for the intensity $|E_0(x, y)|^2$*

7.3.3. *Sideband holography for large amplitude of vibration*

In the previous section (section 7.3.2 and Figure 7.11), we have shown how the comparison of dark fringes for different sidebands leads to a determination of the vibration amplitude $\Phi(x, y)$ at each point of the object. This determination is non-local, since it involves counting fringes from one reference point of the image to the point of interest, so that large amplitudes are not accessible. The vibration amplitude $\Phi(x, y)$ can be determined by another approach that completely eliminates the necessity of counting fringes, giving a local measurement of the amplitude of vibration Φ, even for large Φ [JOU 09b].

For large amplitude of vibration ($\Phi \gg 1$), the distribution of the sideband energy $|E_m|^2$ over m exhibits a sharp variation from maximum to zero near $m \simeq \Phi$, as shown in Figure 7.13 that plots $|E_m|^2/|E|^2$ for $\Phi = 30.3$. This property can be understood if we consider the limits $z_{max} \gg \lambda$. In this case, we can define an instantaneous Doppler angular frequency shift $\omega_D(t)$ and an instantaneous sideband index $m(t)$ which are continuous variables:

$$\omega_D(t) = \omega_A \Phi \cos(\omega_A t) \qquad [7.19]$$

$$m(t) = \frac{\omega_D(t)}{\omega_A} = \Phi \cos(\omega_A t)$$

Because of its sinusoidal variation, $m(t)$ spends more time near the extreme points $m = \pm\Phi$. The Doppler continuous distribution spectrum of $m(t)$ which is displayed in light gray shading in Figure 7.13 are thus maximum near the extreme points $m = \pm\Phi$.

The vibration amplitude $\Phi(x, y)$ can be determined for each location x, y by measuring $|E_m(x, y)|^2$ for all sidebands m, and by determining for each location x, y the sideband index m that corresponds to a fast drop of the signal $|E_m(x, y)|^2$. The method is robust and can easily be used even when the fringes become so narrow that they cannot be resolved, which gives immediate access to large amplitudes of vibration.

By successively adjusting the frequency ω_{LO} of the LO to appropriate values, we record the intensity images $|E_m(x, y)|^2$ of the sidebands as a function of x, y and m. We then obtain a cube of data with three axes x, y and m. Figure 7.14 shows the images obtained for $m =$ 0, 20 ...120 that correspond to cuts of the cube along x, y planes. The images illustrate how, when m increases, the fringes move toward regions with larger amplitudes of vibrations. Since the right part of the reed ($x > 800$) is clamped on the mouthpiece, no signal is obtained in regions near the clamp where $\Phi = 4\pi z_{max}/\lambda \leq m$ (right part of the images).

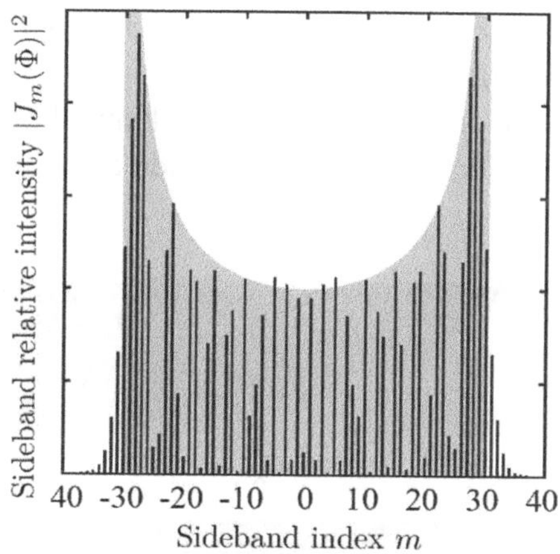

Figure 7.13. *Relative intensity* $|E_m|^2/|E|2 = |J_m(\Phi)|^2$ *of the sideband component* m *for an amplitude modulation of the phase equal to* $\Phi = 30.3$ *radiant. The light gray shading shows the Doppler spectrum obtained from the vibration velocity distribution, with a continuous variable* m*. Both spectra fall abruptly beyond* $m = 30.3$*, which corresponds to the Doppler shift* $\pm\omega_A\Phi$ *associated with the maximum velocity*

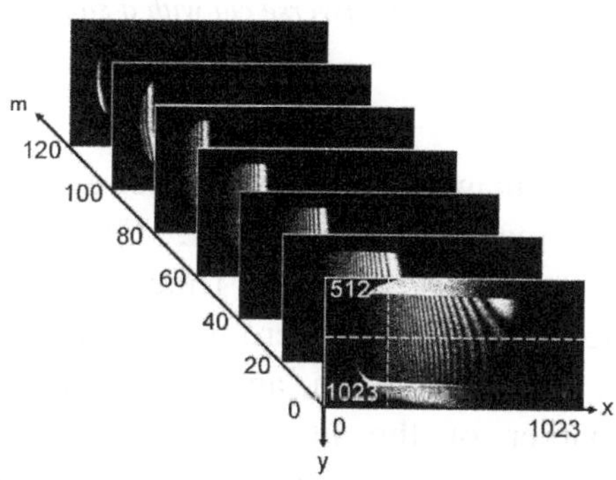

Figure 7.14. *Cube of data obtained from the reconstructed holographic images of a vibrating clarinet reed. The sideband images with* $m = 0, 20, 40 ... 120$ *are shown in an arbitrary linear scale. The white dashed lines correspond to* $x = 249$ *and* $y = 750$*, i.e. to the point chosen for Figure 7.15*

Figure 7.15(a) displays a 2D cut (coordinates x, m) of the cube of data along the horizontal plane $y = 750$ (horizontal white dashed line in Figure 7.14). The envelope of the non-zero (non-black) part of the image provides a measurement of the amplitude of vibration in units of $\lambda/4\pi$. We actually obtain a direct visualization of the shape of the reed at maximal elongation, from the right part clamped on the mouthpiece to the tip on the left. The maximum amplitude corresponds to $\Phi = 120$ rad.

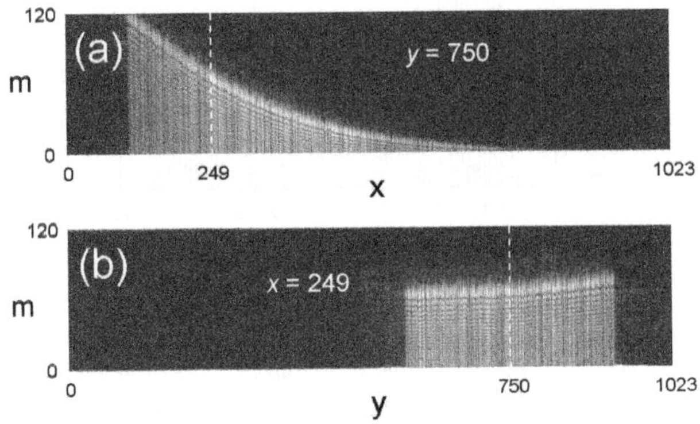

Figure 7.15. *Images corresponding to cuts of 3D data of the reconstructed images along the planes $y = 750$ (a) and $x = 249$ (b). Figure a) shows the deformation of the object along its axis and Figure b) a transverse cut with a slight vibration asymmetry. A logarithmic intensity scale is used*

Figure 7.16 shows images obtained at higher excitation amplitudes, about 10 times larger than that of Figure 7.15. Figure 7.16(a) shows the images obtained for $m = 330$: the fringes are now completely unresolved, but the transition from zero to non-zero intensity remains very clear. With a single hologram, and without fringe counting, we obtain a clear marker of the line where $\Phi(x, y) = 330$ rad. Figure 7.16(b) shows the equivalent of Figure 7.15(a), but with a higher excitation level, and this time for positive and negative values of m. Data range up to about $|m| \simeq 1140$, corresponding to $z_{max} \simeq 58.4$ μm. Since the vibration amplitude is much larger than λ, the continuous approximation for m is valid, and the images of Figure 7.16 can be reinterpreted in terms of classical Doppler effect.

Taking advantage of the sideband order m of the light scattered by a vibrating object adds a new dimension to digital holography. Each pixel x, y of the image then provides an information which is completely independent of others, which results in redundancy and robustness of the measurements. Looking at the edges of the spectrum provides us an accurate determination of the vibration amplitude and avoids a cumbersome analysis of the whole cube of data, thereby giving easy access to measurements of large amplitudes of vibration.

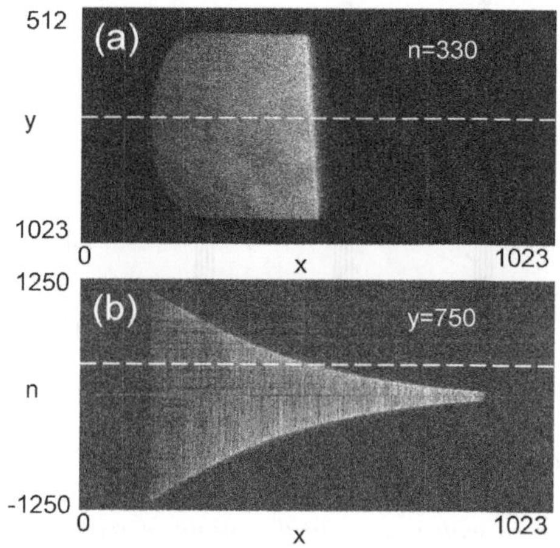

Figure 7.16. *a): image reconstructed with sideband $n = 330$, with a large amplitude of vibration. b) is the equivalent of Figure 7.15(a), but with positive and negative m values. We measure a maximum vibration amplitude of $z_{max} \simeq 60$ μm. A logarithmic intensity scale is used*

7.3.4. *Sideband holography with strobe illumination*

Both time-averaged [PED 95] and sideband digital holography [JOU 09a] record the holographic signal over a large number of vibration periods. These two techniques are not sensitive to the phase of the vibration, and are thus unable to measure the instantaneous velocities of the object. To respond to this problem, Leval *et al.* [LEV 05] combined time-averaged holography with stroboscopic illumination, but, since Leval uses a mechanical stroboscopic device, the Leval technique suffers from a quite low duty cycle (1/144) and is limited in low vibration frequencies ($\omega_A/2\pi < 5$ kHz).

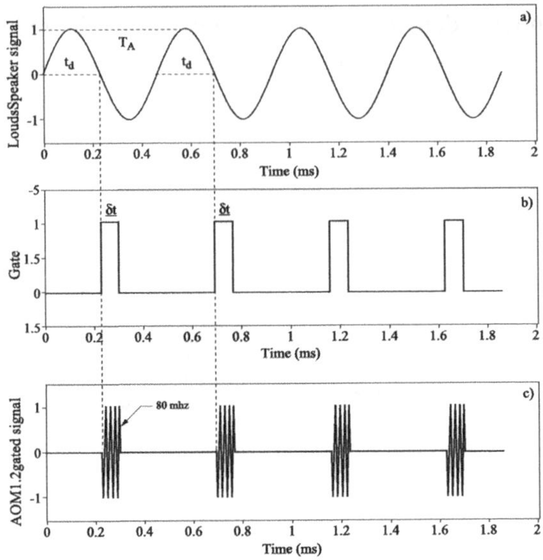

Figure 7.17. *Chronogram of the signals. a) Sinusoidal signal of period $T_A = 2\pi/\omega_A$ exciting the reed. b) Rectangular gate that is applied to both illumination and reference beams. Gate delay is t_d with respect to the origin of reed sinusoidal motion. Gate duration is δt. Gated AOM1 and AOM2 signals at $\simeq$ 80 MHz c). These signals drive the acousto- optic modulators AOM1 and AOM2 and switch on and off the illumination and reference beams*

To overcome these two limitations, it is possible to combine sideband digital holography with stroboscopic illumination and detection synchronized with the vibration [VER 12]. This can be achieved without changing the experimental sideband holography setup of Figure 7.8 by switching electronically on and off the radio frequency (RF) signals which drive the AOMs at ω_{AOM1} and $\omega_{AOM2} \simeq$ 80 MHz. Figure 7.17 shows a typical chronogram of stroboscopic illumination and detection.

Without stroboscopic illumination (and detection), the Doppler velocity spectrum is similar to the spectrum observed in Figure 7.13. It covers the entire range of speeds that can be reached during the sinusoidal motion, that is $\Phi < m < +\phi$, with two maxima near $m \simeq \pm\Phi$. With stroboscopic illumination, the Doppler spectrum is modified, since it corresponds to the narrower velocities range that is reached during gated illumination. However, the finite gate duration δt (that is much shorter than the vibration period $T_A = 2\pi/\omega_A$) yields a broadening of the spectrum of about $T_A/\delta t$ for m. In order to quantitatively take these effects into account, we have calculated the spectrum. The slowly varying complex field $E(t)$ must be multiplied by a gate function $H(t, t_d, \delta t)$ of period T_A. Within the interval $[0, T_A]$, we have:

$$\begin{aligned} H(t, t_d, \delta t) &= 0 \quad \text{for} \quad 0 < t < t_d - \delta t/2 \\ &= 1 \quad \text{for} \quad t_d - \delta t/2 < t < t_d + \delta t/2 \\ &= 0 \quad \text{for} \quad t_d + \delta t/2 < t < T_A \end{aligned} \qquad [7.20]$$

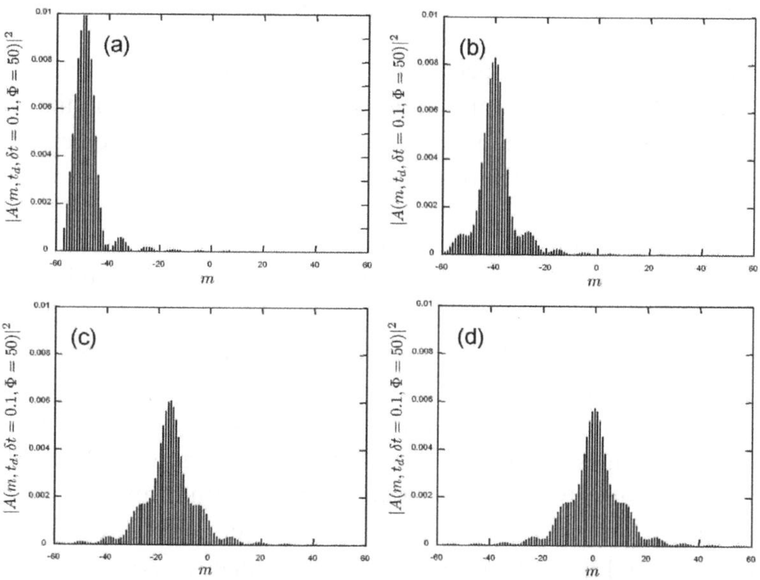

Figure 7.18. *Doppler spectrum $|E_m|^2/|E|^2 = |A(m, t_d, \delta t)|^2$ calculated for a vibration amplitude $\Phi = 0.50$, a gate width $\delta t = 0.1\ T_A$ and a gate time $t_c = 0.25$ a), 0.35 b) , 0.45 c) and 0.50 T_A d)*

Thus, the slowly varying complex field signal $E(t)$ becomes:

$$\begin{aligned} E(t) &= E\, H(t, t_d, \delta t)\; e^{j(\Phi \sin \omega_A t)} \\ &= E \sum_m A(m, t_d, \delta t, \Phi)\; e^{jm\omega_A t} \\ E_m &= E\, A(m, t_d, \delta t, \Phi) \end{aligned} \qquad [7.21]$$

where $A(m, t_d, \delta t)$ is the mth Fourier component of the periodic function $E(t)/E$. We have calculated $A(m, t_d, \delta t, \Phi)$ by Fourier transformation of $E(t)/E$, and have shown in Figure 7.18 the Doppler spectrum $|E_m|^2/|E|^2 = |A(m, t_d, \delta t)|^2$ for different gate times t_c and for $\delta t/T_A = 0.1$. In Figure 7.18(a), the gate coincides with the maximum velocity ($t_d = 0.5\, T_A$). The Doppler spectrum is narrow and centered near $m \simeq -\Phi = -50$ rad. In Figures 7.18(b) and (c), i.e. for gate time $t_c = 0.25$ (a), 0.35 (b), the absolute velocity decreases, and the center of the Doppler spectrum moves toward $m = 0$. On the other hand, the Doppler spectrum broadens. In Figure 7.18(c), the gate coincides with the minimum velocity ($t_d = 0.5\, T_A$). The center of the spectrum is $m = 0$, and the broadening is maximum.

To measure the instantaneous velocities of the object at time $t = t_d$, the gate is adjusted at t_d, and the holograms of all sidebands m are recorded. This can be done automatically by the computer by adjusting the frequency ω_{LO} to fulfill equation 7.18. The sideband images $E_m(x, y)$ are then reconstructed for all m, and a 3D cube of data (coordinates x, y and m) is obtained as shown in Figure 7.14. We can get velocity images by making cuts along x or y. Figure 7.19 shows an example of velocity images obtained at different times t_d of the vibration period. Here, the reed is oriented in the same direction as in Figure 7.14, and the velocity images of coordinates x, m are obtained with horizontal cuts in plane $y = 256$.

7.3.5. *Sideband holography for small amplitude of vibration*

When the vibration amplitude Φ becomes small, the energy within sidebands decreases very rapidly with the sideband indices m, and we only have to consider the carrier $m = 0$, and the two first sidebands $m = \pm 1$. Time-averaged holography [PED 95] that detects the carrier field E_0 is not efficient in detecting small amplitude vibration Φ, because E_0 varies quadratically with Φ:

$$\begin{aligned} E_0(\Phi) &= E\ J_0(\Phi) \\ &\simeq E\,(1 - \Phi^2)/6 \end{aligned} \qquad [7.22]$$

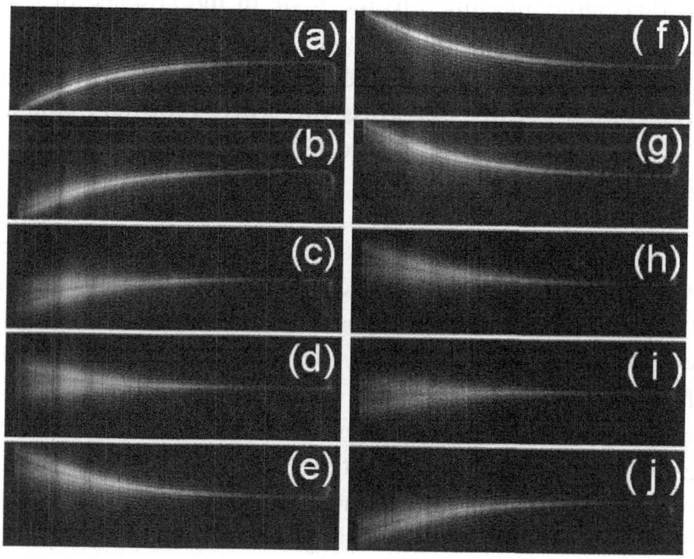

Figure 7.19. *Successive velocities of the reed on a period T_A. These images are obtained by making a cut in the $y = 256$ plane of the 3D stack of reconstructed images, whose coordinates are x, y and m with $m = -100$ to $+100$. Gate duration is $\delta t = 0.1\, T_A$. From a) to j), the gate time t_d is swept from $t_d = 0.25\, T_A$ to $t_d = 1.15\, T_A$ by 0.1 T_A step. The images are displayed in logarithmic scale for the sideband complex field intensity: $|E_m(x, y = 256)|^2$*

However, sideband holography that is able to detect the two first sideband fields $E_{\pm 1}$ is much more sensitive because $E_{\pm 1}$ varies linearly with Φ:

$$\begin{aligned} E_{\pm 1}(\Phi) &= E \; J_{\pm 1}(\Phi) \\ &\simeq \pm E \; \Phi/2 \end{aligned} \quad [7.23]$$

This point was noted about 40 years ago by Ueda *et al.* [UED 76], who did a comprehensive study of the SNR observed for classical (photographic film) sideband holography. The authors managed to observe vibration amplitudes of a few Angstroms, and linked the smallest detectable amplitude to the SNR in the absence of spurious effects. Later on, sub-nanometric vibration amplitude measurements were achieved with sideband digital holography [POW 65, VER 13, BRU 14a, BRU 14c, BRU 14b], and comparison with single point laser interferometry has been made [PSO 12b, BRU 14c].

We must notice that the complex field amplitude E scattered by the sample without vibration strongly depends on the x y position. In a typical experiment, the sample rugosity is such that E is a speckle that is fully developed. The field $E(x, y)$ is thus random in amplitude and phase from one speckle to another. Thus, we cannot extract the phase of the vibration motion from a measurement done on a single sideband, for instance the sideband $m = 1$. To get the phase, we need to measure the field components on several sidebands, for example the carrier field E_0 and the first sideband E_1. For small amplitude Φ, the carrier field $E_0(\Phi)$ is very close to the field $E = E_0(\Phi = 0)$ measured without vibration. By measuring both $E_0(\Phi)$ and $E_1(\Phi)$ by sideband digital holography, we can eliminate E in equations [7.22] and [7.23] getting by the way $J_0(\Phi)$, $J_1(\Phi)$ and Φ. This can be done by calculating either the ratio E_1/E_0 [VER 13]:

$$\begin{aligned} \left| \frac{E_1}{E_0} \right| &= \frac{J_1(\Phi)}{J_0(\Phi)} \\ &\simeq \Phi/2 \end{aligned} \quad [7.24]$$

or the correlation $E_1\ E_0^*$:

$$|E_1\ E_0^*| = |E|^2\ J_1(\Phi)\ J_0(\Phi) \simeq |E|^2\Phi/2 \quad [7.25]$$

Both methods yield a signal ($\Phi/2$ or $|E|^2\Phi/2$) that is proportional to Φ.

From a practical point of view, E_0 and E_1 are measured successively in a standard sideband holography setup like the one in Figure 7.8 [JOU 09a, POW 65]. A sequence of n_{max} frames with $\omega_{LO} = \omega_0 - \omega_{CCD}/4$ is first recorded yielding E_0. A second sequence with $\omega_{LO} = \omega_0 + \omega_A - \omega_{CCD}/4$ is then recorded yielding E_1.

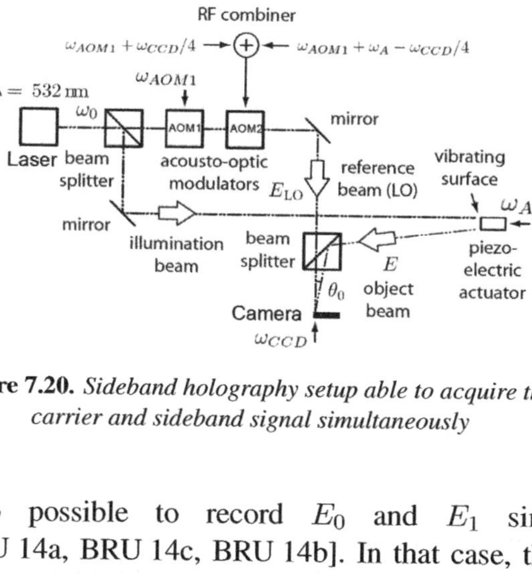

Figure 7.20. *Sideband holography setup able to acquire the carrier and sideband signal simultaneously*

It is also possible to record E_0 and E_1 simultaneously [VER 13, BRU 14a, BRU 14c, BRU 14b]. In that case, the RF signal that drives the second acousto-optic modulator AOM2 is made up of two frequency components (Figure 7.20) at ω^a_{AOM2} and ω^b_{AOM2}:

$$\omega^a_{AOM2} = \omega_{AOM1} + \omega_{CCD}/4 \qquad \omega^b_{AOM2} = \omega_{AOM1} + \omega_A - \omega_{CCD}/4 \quad [7.26]$$

– The first component at ω^a_{AOM2}, whose weight is α, yields the carrier signal E_0, if the hologram H is calculated with demodulation on the -1 grating order, i.e. with $H = \sum_{n=0}^{n_{max}} (-j)^n I_n$.

– The second component at ω^a_{AOM2}, whose weight is β, yields the sideband E_1 with demodulation on the $+1$ grating order, i.e. with $H = \sum_{n=0}^{n_{max}} (+j)^n I_n$.

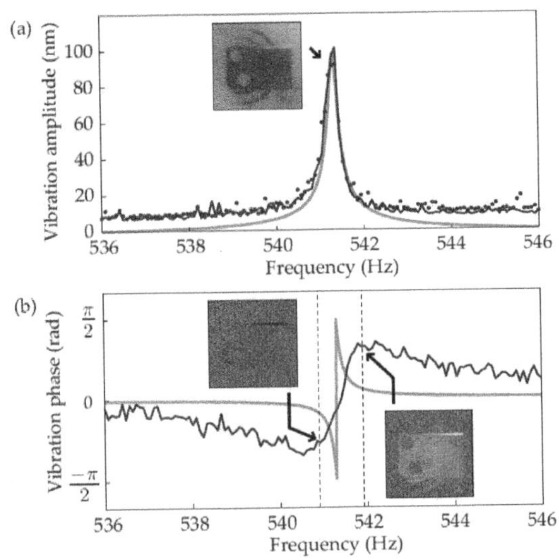

Figure 7.21. *Vibration amplitude z_{max} a) and phase φ b) of a lamellophone, averaged over the first upper cantilever, versus excitation frequency $\omega_A/(2\pi)$ (Hz). Insets: retrieved vibration amplitude and phase maps in the neighborhood of the resonance. The theoretical resonance line is shown in gray*

Simultaneous detection of E_0 and E_1, and calculation of Φ by the ratio method of equation 7.24, has been used to study various vibrating samples. Figure 7.21 shows an example of signals obtained with a lamellophone of a musical box [BRU 14a]. In that experiment, all the frequencies ω_A, ω_{CCD}, ω_{AOM1}, ω^a_{AOM2} and ω^b_{AOM2} are driven by numerical synthesizers. In this particular case, harmonics are extracted by a temporal Fourier transform of a sequence of eight raw interferograms. Here, optical sidebands of interest are downconverted to demodulated frequencies in the camera bandwidth. In order to avoid signal phase drifts from the measurement of one sequence to the next, AOM2 is driven using a phase ramp. As a result, the phases of E_0 and E_1 are well defined, and yield the phase φ of the mechanical motion. We have:

$$\varphi = \text{Arg}(E_1/E_0) + C \qquad [7.27]$$

where C is an additive constant that depends of the relative phase of the synthesizers. Figure 7.21 shows the averaged amplitude z_{max} and phase φ of the upper cantilever of the lamellophone, which is driven through its resonance frequency $\simeq 541.5$ Hz. As expected, the phase φ makes a jump of about π when excitation frequency ω_A crosses the resonance frequency.

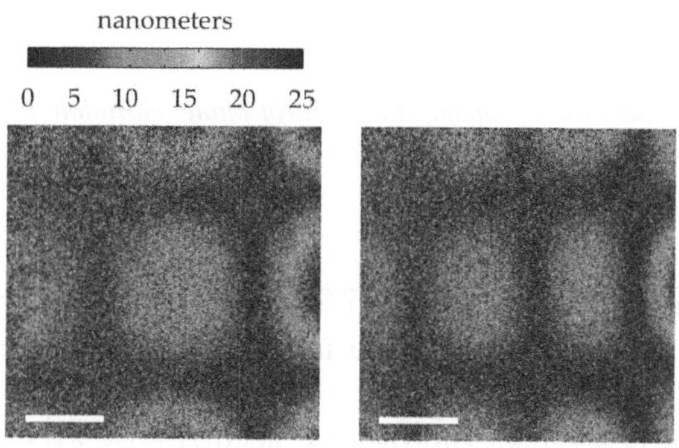

Figure 7.22. *Amplitude maps of the out-of-plane vibration of a thin metal plate versus excitation frequency $\omega_A/(2\pi)$. Holographic images at 40.1 kHz a), 61.7 kHz b). Scalebar: 5 mm*

The method has also been applied to get full-field images of surface acoustic waves as shown in Figure 7.22 [BRU 14c]. Comparison with single-point Doppler vibrometry is shown in Figure 7.23. If the carrier and sideband components of the RF signal at ω^a_{AOM2} and ω^b_{AOM2} have the same amplitude (i.e. if $\alpha = \beta$), some spurious detection of the carrier field E_0 (which is much bigger) is observed when the detection is tuned on the grating order -1 (used to detect E_0). This spurious effect limits the detection sensitivity to $z_{max} \sim 1$ nm, as shown in Figure 7.22 ($\circ$ symbols). Much better results are obtained by reducing the weight of the carrier component (i.e. with $\beta \gg \alpha$). The detection sensitivity becomes $z_{max} \sim 0.01$ nm with $\beta \simeq 50\alpha$ as shown in Figure 7.22 (+ symbols).

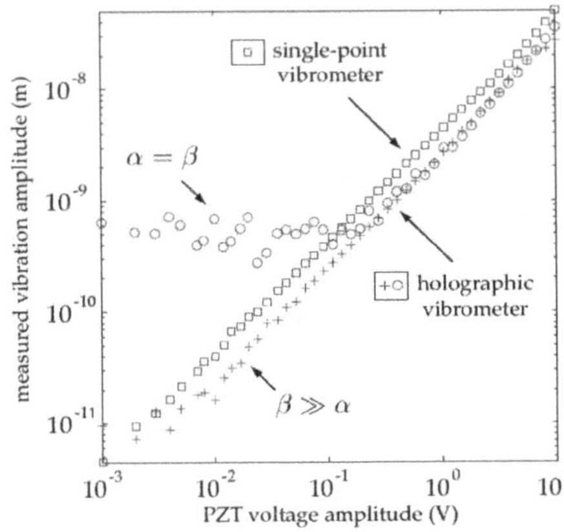

Figure 7.23. *Comparison of quantitative out-of-plane vibration amplitudes z_{max} retrieved with the single-point laser vibrometer (□ symbols) and holographic vibrometer with $\beta = \alpha$ (○ symbols) and $\beta \simeq 50\alpha$ (+ symbols)*

The ratio method of equation [7.24] used above is simple, since it directly gives Φ. Nevertheless, the method is unstable for the points x, y of the object where E is close to zero. Since Φ varies slowly with the location x, y, we can improve the accuracy by averaging Φ over neighbor points. In that case, the ratio method is not optimal for noise, since all points x, y are weighted equally in the average, while the accuracy of the ratio strongly depends on the location x, y, since the accuracy is low when $|E(x, y)|$ is low.

For this problem, the correlation method of equation [7.25] is less simple, since it gives $|E|^2\Phi$, and not Φ. It is, nevertheless, better for noise, since spatial averaging over neighbor points x, y is weighted by the scattered energy $|E(x, y)|^2$.

Figure 7.24 shows reconstructed images of the correlation $E_1 E_0^*$ obtained with a cube of wood (2 cm $\times$ 2 cm) vibrating at its resonance frequency $\omega_A = 21.43$ kHz. Here, the fields E_0 and E_1 are measured successively. In order to get better SNR, the complex correlation signal

$E_1(x,y)E_0^*(x,y)$ is averaged over neighbor x, y points by using a 2D Gaussian blur filter of radius 4 pixels. Figure 7.24 shows (a) the amplitude and (b) the phase of the filtered correlation $E_1(x,y)E_0^*(x,y)$. Figure 7.24(c) displays the phase of the correlation in 3D. As can be seen in the figure, the opposite corners (upper left and bottom right, for example) vibrate in phase, while the neighbor corners (upper left and upper right, for example) vibrate in phase opposition. Note that the cube is excited in one of its corners by a needle. This may explain why the opposite corners are not perfectly in phase in Figures 7.24(b) and (c).

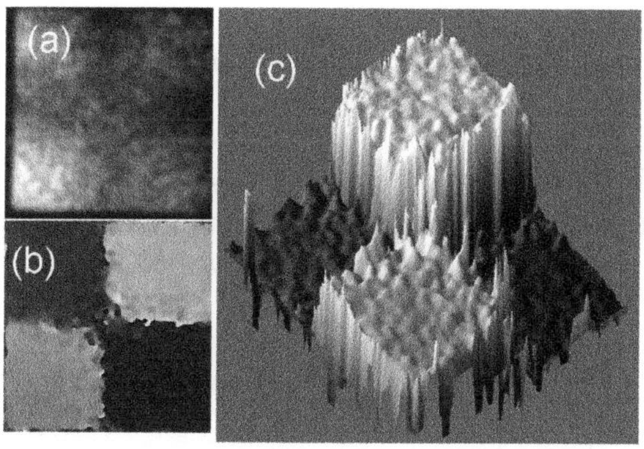

Figure 7.24. *(a,b) Reconstructed images of the correlation $E_1 E_0^*$ obtained with a cube of wood; a) amplitude: $|E_1 E_0^*|$, and b) phase: $\arg E_1 E_0^*$. The correlation $E_1(x,y)E_0^*(x,y)$ is averaged with a 2D Gaussian blur filter of radius 4 pixels. The display is made in arbitrary linear scale. c) 3D display of the phase: $x, y, \arg E_1 E_0^*$*

In order to increase the ability to analyze vibrations of small amplitude, we can increase the illumination on the zone where the correlation averaging is performed. This is done by using a collimated illumination beam. Figure 7.25 shows an example. The illumination laser has been focused on point x_0, y_0 of the clarinet reed to be studied (see Figure 7.25(a)). Most of the energy in the reconstructed image is thus located near x_0, y_0 (see Figure 7.25(b)). To measure the vibration amplitude at location x_0, y_0, we have calculated the averaged correlation $\langle E_1 E_0^* \rangle$ and the averaged energy $\langle |E_0|^2 \rangle$:

$$\begin{aligned}\langle E_1 E_0^* \rangle &= (1/N_{pix}) \sum_{x,y} E_1(x,y) E_0^*(x,y) \\ &= J_1(\Phi) J_0(\Phi)\ (1/N_{pix}) \sum_{x,y} |E(x,y)|^2\end{aligned} \quad [7.28]$$

$$\begin{aligned}\langle |E_0|^2 \rangle &= (1/N_{pix}) \sum_{x,y} |E_0(x,y)|^2 \\ &= J_0^2(\Phi)\ (1/N_{pix}) \sum_{x,y} |E(x,y)|^2\end{aligned} \quad [7.29]$$

where $\sum_{x,y}$ is the summation over the N_{pix} pixels of the illuminated region located near x_0, y_0. In that region, the vibration amplitude Φ is supposed not to depend on x and y. Then, we get:

$$\frac{\langle E_1 E_0^* \rangle}{\langle |E_0|^2 \rangle} = \frac{J_1(\Phi)}{J_0(\Phi)} \simeq \Phi/2 \quad [7.30]$$

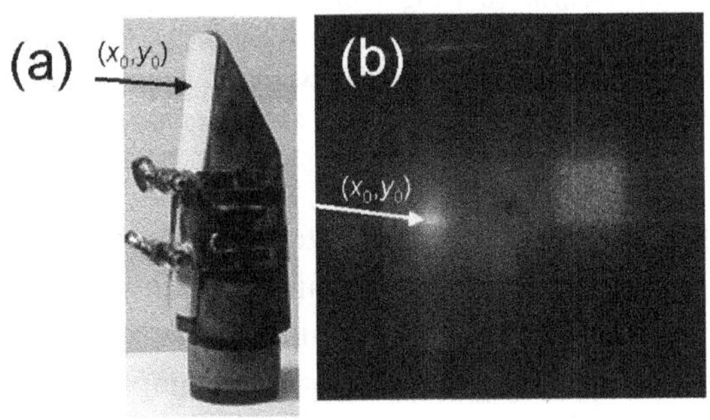

Figure 7.25. *a) Clarinet reed with illumination beam focused in x_0, y_0. b) Sideband $m = 1$ reconstructed image of the vibrating reed. The display is made in arbitrary log scale for the field intensity $|E_1(x,y)|^2$*

To verify the ability of measuring small vibration, the reed has been excited with increasingly lower loudspeaker levels, the illumination laser being focused on the point x_0, y_0 of Figure 7.25(b), we want to study. A sequence of $n_{max} = 128$ frames have been recorded for both carrier (E_0) and sideband (E_1). Then four-phase detection has been done by calculating H with $n_{nmax} = 128$ in equation [7.9]. Then, we have calculated the reconstructed fields $E_0(x, y)$ and $E_1(x, y)$, the averaged correlation $\langle E_1 E_0^* \rangle$ and averaged carrier intensity $\langle |E_0|^2 \rangle$, and the ratio $\langle E_1 E_0^* \rangle / \langle |E_0|^2 \rangle$. This ratio is shown in Figure 7.26 (y-axis) as a function of the peak-to-peak voltage V_{pp} of the sinusoidal signal that excites the loudspeaker (x-axis). The measured ratio varies linearly with V_{pp}. For low voltage $V_{pp} \leq 5$ mV, the ratio reached a noise floor corresponding to $\langle E_1 E_0^* \rangle / \langle |E_0|^2 \rangle \simeq 10^{-4}$. Since this voltage V_{pp} is proportional to the vibration amplitude z_{max}, and thus to Φ, we have converted the voltage V_{pp} into vibration amplitude z_{max}. The conversion factor ($z_{max} = 0.02$ nm for $V_{pp} = 10$ mV) is obtained by assuming that $J_1(\Phi)/J_0(\Phi)$ fits the ratio $\langle E_1 E_0^* \rangle / \langle |E_0|^2 \rangle$ that is measured, as predicted by equation [7.30] (see the solid gray curve of Figure 7.26). We get here a noise floor corresponding to $z_{max} \simeq 0.01$ nm. It is about 10 times lower than the limit $\lambda/3,500 = 0.22$ nm predicted by Ueda *et al.* [UED 76], and 10 times lower than in previous experiments at similar frequency [POW 65, VER 13] in the kHz range. Similar noise floor 0.01 nm has been obtained by the ratio method in Figure 7.23, but at a higher vibration frequency $\omega \simeq 40$ kHz [BRU 14c].

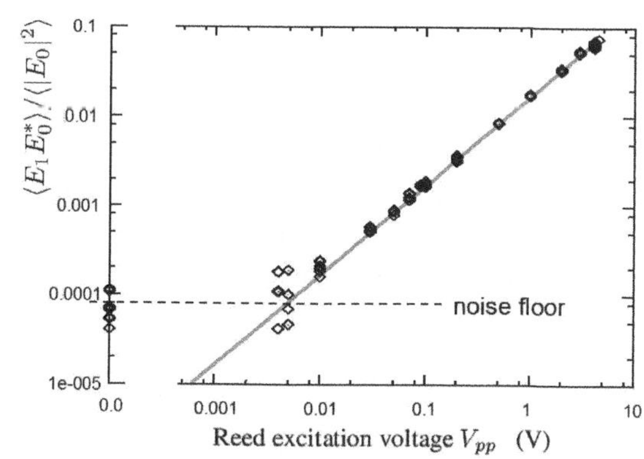

Figure 7.26. *Ratio $\langle E_1 E_0^* \rangle / \langle |E_0|^2 \rangle$ (vertical axis) as a function of the reed excitation voltage V_{pp} in volt units. The light gray curve is $J_1(\Phi)/J_0(\Phi)$ with $\Phi = z_{max}/(4\pi\lambda)$*

We have tried to make the experiment with more frames, i.e. with $n_{max} > 128$, but this does not lower the noise floor. The remaining noise floor can be related to a spurious detection of the carrier field E_0 when detection is tuned to detect E_1, or to some technical noise on the carrier signal that is not totally filtered off. Since the carrier and sideband fields E_0 and E_1 are within the same spatial mode, no space filtering can be applied to filter off E_0. Here, the spurious carrier field E_0 is filtered off in the time domain, while the LO field E_{LO} is filtered off both in time and frequency domains.

7.4. Conclusion

In this chapter, we have presented the digital heterodyne holography technique that is able to fully control the amplitude, phase and frequency of both illumination and reference beams. Full automatic data acquisition of the holographic signal can be made possible, and ultimate shot-noise sensitivity can be achieved. Heterodyne holography is an extremely versatile and powerful tool, particularly when applied to vibration analysis. In this case, heterodyne holography is able to detect in wide field (i.e. in all points of the object 2D surface at the same time) the vibrating object signal at any optical sideband of rank m. Since the control of the intensity of the illumination and reference beams is fast, instantaneous measurements of the vibration signal, sensitive to the mechanical phase, can be done. For the measurement of large vibration amplitudes, the possible ambiguity of measurements can be removed by making measurements at different sideband indices m. For small vibration amplitudes, the mechanical phase can be obtained from measurements done on the carrier ($m = 0$) and on the first sideband ($m = 1$). Moreover, extremely low vibration amplitudes, below 10 pm, can be measured.

7.5. Bibliography

[ATL 07] ATLAN M., GROSS M., ABSIL E., "Accurate phase-shifting digital interferometry", *Optics Letters*, vol. 32, no. 11, pp. 1456–1458, 2007.

[BRU 14a] BRUNO F., LAUDEREAU J.-B., LESAFFRE M., *et al.*, "Phase-sensitive narrowband heterodyne holography", *Applied Optics*, vol. 53, no. 7, pp. 1252–1257, 2014.

[BRU 14b] BRUNO F., LAURENT J., PRADA C., *et al.*, "Non-destructive testing of composite plates by holographic vibrometry", *Journal of Applied Physics*, vol. 115, no. 15, p. 154503, 2014.

[BRU 14c] BRUNO F., LAURENT J., ROYER D., *et al.*, "Holographic imaging of surface acoustic waves", *Applied Physics Letters*, vol. 104, no. 8, p. 083504, 2014.

[CUC 00] CUCHE E., MARQUET P., DEPEURSINGE C., "Spatial filtering for zero-order and twin-image elimination in digital off-axis holography", *Applied Optics*, vol. 39, no. 23, pp. 4070–4075, 2000.

[DEM 04] DEMOLI N., VUKICEVIC D., "Detection of hidden stationary deformations of vibrating surfaces by use of time-averaged digital holographic interferometry", *Optics Letters*, vol. 29, no. 20, pp. 2423–2425, 2004.

[FU 07] FU Y., PEDRINI G., OSTEN W., "Vibration measurement by temporal Fourier analyses of a digital hologram sequence", *Applied Optics*, vol. 46, no. 23, pp. 5719–5727, 2007.

[GRO 07] GROSS M., ATLAN M., "Digital holography with ultimate sensitivity", *Optics Letters*, vol. 32, no. 8, pp. 909–911, 2007.

[GRO 08] GROSS M., ATLAN M., ABSIL E., "Noise and aliases in off-axis and phase-shifting holography", *Applied Optics*, vol. 47, no. 11, pp. 1757–1766, 2008.

[JOU 09a] JOUD F., LALOË F., ATLAN M., *et al.*, "Imaging a vibrating object by Sideband Digital Holography", *Optics Express*, vol. 17, no. 4, pp. 2774–2779, 2009.

[JOU 09b] JOUD F., VERPILLAT F., LALOË F., *et al.*, "Fringe-free holographic measurements of large-amplitude vibrations", *Optics letters*, vol. 34, no. 23, pp. 3698–3700, 2009.

[LE 00] LE CLERC F., COLLOT L., GROSS M., "Numerical heterodyne holography with two-dimensional photodetector arrays", *Optics Letters*, vol. 25, no. 10, pp. 716–718, 2000.

[LE 01] LE CLERC F., GROSS M., COLLOT L., "Synthetic-aperture experiment in the visible with on-axis digital heterodyne holography", *Optics Letters*, vol. 26, no. 20, pp. 1550–1552, 2001.

[LES 13] LESAFFRE M., VERRIER N., GROSS M., "Noise and Signal scaling factors in Digital Holography in week illumination: relationship with Shot Noise", *Applied Optics*, vol. 52, pp. A81–A91, 2013.

[LEV 05] LEVAL J., PICART P., BOILEAU J.P., *et al.*, "Full-field vibrometry with digital Fresnel holography", *Applied Optics*, vol. 44, no. 27, pp. 5763–5772, 2005.

[PED 98] PEDRINI G., FROENING P., FESSLER H., *et al.*, "Transient vibration measurements using multi-pulse digital holography", *Optics & Laser Technology*, vol. 29, no. 8, pp. 505–511, 1998.

[PED 06] PEDRINI G., OSTEN W., GUSEV M.E., "High-speed digital holographic interferometry for vibration measurement", *Applied Optics*, vol. 45, no. 15, pp. 3456–3462, 2006.

[PED 95] PEDRINI G., ZOU Y., TIZIANI H., "Digital double-pulsed holographic interferometry for vibration analysis", *Journal of Modern Optics*, vol. 42, no. 2, pp. 367–374, 1995.

[PED 97] PEDRINI G., TIZIANI H., ZOU Y., "Digital double pulse-TV-holography", *Optics and Lasers in Engineering*, vol. 26, no. 2, pp. 199–219, 1997.

[PIC 03] PICART P., LEVAL J., MOUNIER D., *et al.*, "Time-averaged digital holography", *Optics Letters*, vol. 28, no. 20, pp. 1900–1902, 2003.

[PIC 05a] PICART P., LEVAL J., MOUNIER D., *et al.*, "Some opportunities for vibration analysis with time averaging in digital Fresnel holography", *Applied Optics*, vol. 44, no. 3, pp. 337–343, 2005.

[PIC 05b] PICART P., LEVAL J., PASCAL J. C., *et al.*, "2D full field vibration analysis with multiplexed digital holograms", *Optics Express*, vol. 13, no. 22, pp. 8882–8892, 2005.

[POW 65] POWELL R.L., STETSON K.A., "Interferometric vibration analysis by wavefront reconstruction", *Journal of the Optical Society of America*, vol. 55, no. 12, pp. 1593–1597, 1965.

[PSO 12a] PSOTA P., LEDL V., DOLECEK R., *et al.*, "Measurement of piezoelectric transformer vibrations by digital holography", *IEEE Transactions on Ultrasonics, Ferroelectrics and Frequency Control*, vol. 59, no. 9, pp. 1962–1968, 2012.

[PSO 12b] PSOTA P., LÉDL V., DOLEČEK R., *et al.*, "Comparison of digital holographic method for very small amplitudes measurement with single point laser interferometer and laser Doppler vibrometer", *Digital Holography and Three-Dimensional Imaging*, Optical Society of America, p. DSu5B-3, 2012.

[SCH 94] SCHNARS U., JUPTNER W., "Direct recording of holograms by a CCD target and numerical reconstruction", *Applied Optics*, vol. 33, no. 2, pp. 179–181, 1994.

[TAI 14] TAILLARD P.-A., LALOË F., GROSS M., *et al.*, "Statistical estimation of mechanical parameters of clarinet reeds using experimental and numerical approaches", *Acta Acustica united with Acustica*, vol. 100, no. 3, pp. 555–573, 2014.

[UED 76] UEDA M., MIIDA S., SATO T., "Signal-to-noise ratio and smallest detectable vibration amplitude in frequency-translated holography: an analysis", *Applied Optics*, vol. 15, no. 11, pp. 2690–2694, 1976.

[VER 10] VERPILLAT F., JOUD F., ATLAN M., *et al.*, "Digital holography at shot noise level", *Journal of Display Technology*, vol. 6, no. 10, pp. 455–464, 2010.

[VER 12] VERPILLAT F., JOUD F., ATLAN M., *et al.*, "Imaging velocities of a vibrating object by stroboscopic sideband holography", *Optics Express*, vol. 20, no. 20, pp. 22860–22871, 2012.

[VER 13] VERRIER N., ATLAN M., "Absolute measurement of small-amplitude vibrations by time-averaged heterodyne holography with a dual local oscillator", *Optics Letters*, vol. 38, no. 5, pp. 739–741, 2013.

[YAM 97] YAMAGUCHI I., ZHANG T., "Phase-shifting digital holography", *Optics Letters*, vol. 22, no. 16, pp. 1268–1270, 1997.

List of Authors

Michael ATLAN
Langevin Institute UPMC
UP7, ESPCI ParisTech
Paris
France

Marc BRUNEL
CORIA – University of Rouen
Saint Etienne du Rouvray
France

Sébastien COËTMELLEC
CORIA – University of Rouen
Saint Etienne du Rouvray
France

Jean Michel DESSE
ONERA
Lille
France

Frank DUBOIS
Université Libre de Bruxelles
Brussels
Belgium

Ahmed EL MALLAHI
Université Libre de Bruxelles
Brussels
Belgium

Marc GEORGES
Centre Spatial de Liège
University of Liège
Belgium

Michel GROSS
Charles Coulomb Laboratory
University of Montpellier 2
Montpellier
France

Denis LEBRUN
CORIA – University of Rouen
Saint Etienne du Rouvray
France

Pierre MARQUET
EPFL SV BMI LNDC
Lausanne
Switzerland

Christophe MINETTI
Université Libre de Bruxelles
Brussels
Belgium

Nicolas PAVILLON
Osaka-fu, Minoh-shi
Japan

Pascal PICART
ENSIM
University of Maine
Le Mans
France

Benjamin RAPPAZ
EPFL SV BMI LNDC
Lausanne
Switzerland

Nicolas VERRIER
Laboratoire Hubert Currien
University of St Etienne
France

Index

G, H

I

Q, R, S

T, U, V

W, Z

Other titles from

iSTE

in

Instrumentation and Measurement

2014

APPRIOU Alain
Uncertainty Theories and Multisensor Data Fusion

THOMAS Jean-Hugh, YAAKOUBI Nourdin
New Sensors and Processing Chain

2012

LALAUZE René
Chemical Sensors and Biosensors

LE MENN Marc
Instrumentation and Metrology in Oceanography

2011

GROUS Ammar
Applied Metrology for Manufacturing Engineering

2009

FRENCH COLLEGE OF METROLOGY
Transverse Disciplines in Metrology

2008

LALAUZE René
Physical Chemistry of Solid-Gas Interfaces

2007

KUNDU Tribikram
Advanced Ultrasonic Methods for Material and Structure Inspection

PLACKO Dominique
Fundamentals of Instrumentation and Measurement

RIPKA Pavel, TIPEK Alois
Modern Sensors Handbook

2006

BALAGEAS Daniel *et al.*
Structural Health Monitoring

FRENCH COLLEGE OF METROLOGY
Metrology in Industry

Printed and bound by CPI Group (UK) Ltd, Croydon, CR0 4YY
16/09/2026
14963022-0005